Managing Electrical Hazards

SIXTH EDITION

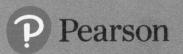 

Pearson

NCCER | National Center for Construction Education and Research

NCCER

President and Chief Executive Officer: Boyd Worsham
Vice President of Innovation and Advancement: Jennifer Wilkerson
Chief Learning Officer: Lisa Strite
Senior Manager, Curriculum Development: Chris Wilson
Production Manager: Graham Hack
Managing Electrical Hazards Project Manager: Douglas Hancock
Technical Writing Manager: Gary Ferguson
Art Manager: Bree Rodriguez
Technical Illustrators: Judd Ivines, Liza Wailes
Production Artist: Chris Kersten
Permissions Specialists: Adam Black, Sherry Davis
Managing Editor: Hannah Murray
Editors: Zi Meng, Alexandria Willbond, James Singer III, Lauren Tygrest
Desktop Publishing Manager: Eric Caraballoso III
Desktop Publishing Coordinator: Daphney Milian
Production Specialist: Julie Watkins
Digital Content Manager: Kelly Beck
Digital Content Coordinator and Translation Manager: Yesenia Tejas
Digital Content Coordinator: Briana Rosa
Project Coordinators: Chelsi Santana, Colleen Duffy, Milagro Maradiaga

Pearson

Director of Association Partnerships: Tanja Eise
Program Manager: Vanessa Price
Senior Digital Content Producer: Shannon Stanton
Content Producer: Arup Kumar Ghosh
Employability Solutions Coordinator: Monica Perez
Cover Designer: Mary Siener
Rights and Permissions: Jenell Forschler

Composition: NCCER
Content Technologies: Gnostyx
Printer/Binder: Lakeside Book Company
Cover Printer: Lakeside Book Company
Text Typefaces: Palatino LT Pro and Helvetica Neue

Cover Image

Cover photo provided by: Oberon

10 9 8 7 6 5 4 3 2 1   12 2024

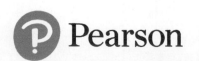

Paperback:
ISBN-13: 978-0-13-831869-7

PREFACE

To the Trainee

Electrical power can be deadly. Nearly everyone who has worked on electrical equipment has experienced electrical shock at some level, and some have suffered shocks that caused serious injury or death. Under certain conditions, electrical equipment can explode, resulting in a blast of extreme heat and shrapnel flying in all directions. Anyone within range of such a blast can be seriously burned or injured by the flying material. This module discusses electrical shock and arc flash risks and explains how to use NFPA 70E®, *Standard for Electrical Safety in the Workplace*®, as a tool for recognizing and managing these hazards. It also covers the use of personal protective equipment (PPE), employer/employee responsibilities for providing and maintaining a safe workplace, and building a personal safety toolbox. This module was designed primarily for safety managers, supervisors, superintendents, and others who may not be directly involved in electrical work but need to understand the requirements and methods used in electrical safety performance. A copy of the 2024 NFPA 70E®, *Standard for Electrical Safety in the Workplace*®, is a prerequisite material for this course. To order, contact NFPA at **www.nfpa.org** or call 1-800-344-3555.

New with *Managing Electrical Hazards*

This edition of *Managing Electrical Hazards* has been revised to align with changes in the updated 2024 NFPA 70E®, *Standard for Electrical Safety in the Workplace*®. Along with aligning the content, we have updated the text, illustrations, and photos to reflect today's approach to electrical safety. The addition of the Know the Code features also alerts students to important code-related references within the text. Lastly, all these changes are presented in a new layout that is visually appealing and easier to read.

We wish you success as you progress through this training program. If you have any comments on how NCCER might improve upon this textbook, please complete the User Update form using the QR code on this page. NCCER appreciates and welcomes its customers' feedback. You may submit yours by emailing **support@nccer.org**. When doing so, please identify feedback on this title by listing *#MEH* in the subject line.

Our website, **www.nccer.org**, has information on the latest product releases and training.

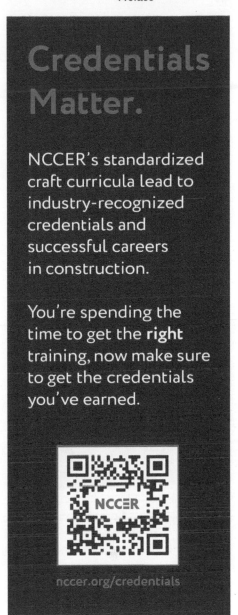

Credentials Matter.

NCCER's standardized craft curricula lead to industry-recognized credentials and successful careers in construction.

You're spending the time to get the **right** training, now make sure to get the credentials you've earned.

nccer.org/credentials

SCAN ME

NCCER Standardized Curricula

NCCER is a not-for-profit 501(c)(3) education foundation established in 1996 by the world's largest and most progressive construction companies and national construction associations. It was founded to address the severe workforce shortage facing the industry and to develop a standardized training process and curricula. Today, NCCER is supported by hundreds of leading construction and maintenance companies, manufacturers, and national associations. The NCCER Standardized Curricula was developed by NCCER in partnership with Pearson, the world's largest educational publisher.

Some features of the NCCER Standardized Curricula are as follows:

- An industry-proven record of success
- Curricula developed by the industry, for the industry
- National standardization providing portability of learned job skills and educational credits
- Compliance with the Office of Apprenticeship requirements for related classroom training (*CFR 29:29*)
- Well-illustrated, up-to-date, and practical information

NCCER maintains a secure online database that provides certificates, digital badges, transcripts, and wallet cards to individuals who successfully complete programs under an NCCER-accredited organization or through one of NCCER's self-paced, online programs. This system also allows individuals and employers to track and verify industry-recognized credentials and certifications in real time.

For information on NCCER's credentials, contact NCCER Customer Service at 1-888-622-3720 or visit **www.nccer.org**.

Digital Credentials

Show off your industry-recognized credentials online with NCCER's digital credentials.

NCCER is now providing online credentials. Transform your knowledge, skills, and achievements into digital credentials that you can share across social media platforms, send to your network, and add to your resume. For more information, visit **www.nccer.org**.

Cover Image

Oberon was founded in 1978 by Jack Hirschmann to offer innovative personal protection products. After development of the Face Fit® Face Shield and their Gold Shield technology, the company pioneered the first Arc Flash Face Shield in 1985. Oberon's dedication to engineering innovative safety products continues today with the development of their True Color Grey (TCG™) Arc Flash Face Shields.

DESIGN FEATURES

Content is organized and presented in a functional structure that allows you to access the information where you need it.

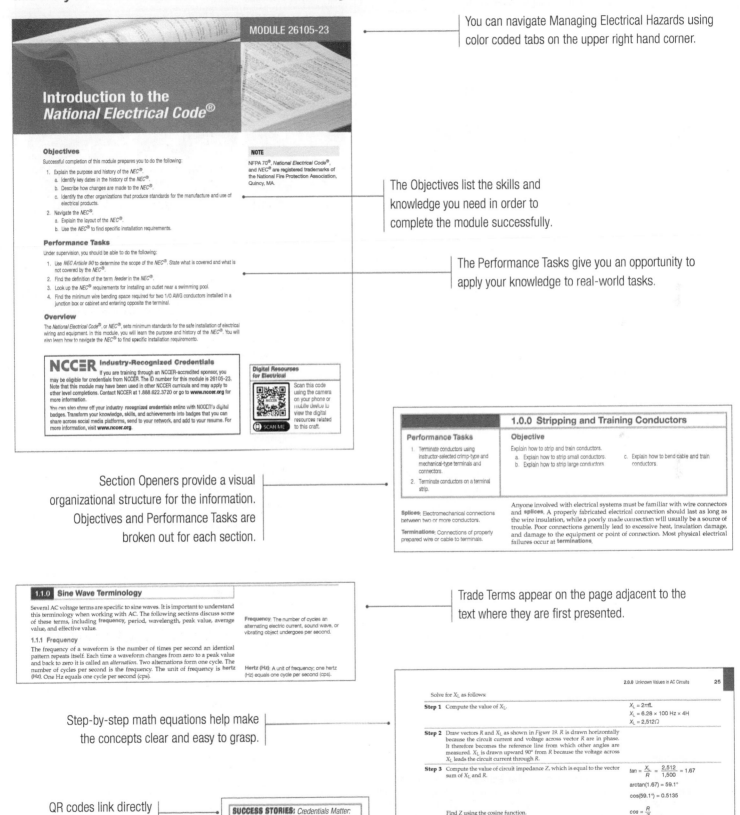

You can navigate Managing Electrical Hazards using color coded tabs on the upper right hand corner.

The Objectives list the skills and knowledge you need in order to complete the module successfully.

The Performance Tasks give you an opportunity to apply your knowledge to real-world tasks.

Section Openers provide a visual organizational structure for the information. Objectives and Performance Tasks are broken out for each section.

Trade Terms appear on the page adjacent to the text where they are first presented.

Step-by-step math equations help make the concepts clear and easy to grasp.

QR codes link directly to videos that highlight current content.

Important information is highlighted, illustrated, and presented to facilitate learning.

Placement of images near the text description and details such as callouts and labels help you absorb information.

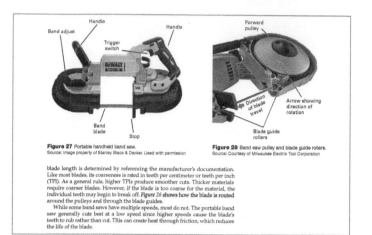

Figure 27 Portable handheld band saw.
Source: Image property of Stanley Black & Decker. Used with permission

Figure 28 Band saw pulley and blade guide rollers.
Source: Courtesy of Milwaukee Electric Tool Corporation

blade length is determined by referencing the manufacturer's documentation. Like most blades, its coarseness is rated in teeth per centimeter or teeth per inch (TPI). As a general rule, higher TPIs produce smoother cuts. Thicker materials require coarser blades. However, if the blade is too coarse for the material, the individual teeth may begin to break off. *Figure 28* shows how the blade is routed around the pulleys and through the blade guides.

While some band saws have multiple speeds, most do not. The portable band saw generally cuts best at a low speed since higher speeds cause the blade's teeth to rub rather than cut. This can create heat through friction, which reduces the life of the blade.

Preparing Drills with Keyless Chucks

Most cordless drills use a keyless chuck. While the steps for preparing a cordless drill are similar, there are some small differences. Follow the steps below when preparing to use drills with keyless chucks:

Step 1 Disconnect the drill from its power source by removing the battery pack before loading a bit.

Step 2 As shown in (*Figure 7A*), open the chuck by turning it counterclockwise until the jaws are wide enough to insert the bit shank.

Step 3 Insert the bit shank into the chuck opening (*Figure 7B*). Keeping the bit centered in the opening, turn the chuck by hand until the jaws grip the bit shank.

Step 4 Tighten the chuck securely with your hand so that the bit does not move (*Figure 7C*). You are now ready to use the cordless drill.

(A) Insert the Bit Shank. **(B) Keep Bit Straight and Partially Tighten the Chuck.** **(C) Tighten the Chuck Securely.**

Figure 7 Loading the bit on a keyless chuck.
Source: Cianbro Corporation

New boxes highlight safety and other important information. Warning boxes stress potentially dangerous situations, while Caution boxes alert to dangers that may cause damage to equipment. Note boxes provide additional information on a topic.

> **WARNING!**
>
> A portable band saw always cuts in the direction of the user. For that reason, workers must be especially careful to avoid injury when using this type of saw. Always wear appropriate PPE and stay focused on the work.

> **CAUTION**
>
> Never assume anything. It never hurts to ask questions, but disaster can result if you don't ask. For example, do not assume that an electrical power source is turned off. First ask whether the power is turned off, then check it yourself to be completely safe.

> **NOTE**
>
> This training alone does not provide any level of certification in the use of fall arrest or fall restraint equipment. Trainees should not assume that the knowledge gained in this module is sufficient to certify them to use fall arrest equipment in the field.

> **Did You Know?**
> **Lightning Rods**
>
> An interesting fact about grounding is that a lightning rod (air terminal) isn't meant to bring a bolt of lightning to ground. To do this, its conductors would have to be several feet (1 m or more) in diameter. The purpose of the rod is to dissipate the negative static charge that would cause the positive lightning charge to strike the house.

These boxed features provide additional information that enhances the text.

> **Around the World**
> **Predicting Weather**
>
> Barometers measure atmospheric air pressure. Generally, an increased pressure indicates that the weather is pleasant. A reduced pressure indicates that the weather is cloudy and/or rainy.
>
> Changes in barometric pressure are used to predict changes in weather conditions. For example, if it is raining outside, but the barometer is rising, it indicates that better weather is on the way. In the midst of a hurricane, the barometric pressure is an indicator of intensity. Extremely low pressures exist in powerful hurricanes. While traditional barometers work very well, sophisticated weather stations capture far more data and can transmit the information wirelessly.

> **Orthographic Drawings**
>
> Orthographic drawings are used for elevation drawings. They show straight-on views of the different sides of an object with dimensions that are proportional to the actual physical dimensions. In orthographic drawings, the designer draws lines that are scaled-down representations of real dimensions. Every 12 inches, for example, may be represented by ¼ inch on the drawing. Similarly, in an example using metric measurements with a ratio of 1:2, every 30 millimeters may be represented by 15 millimeters on the drawing.

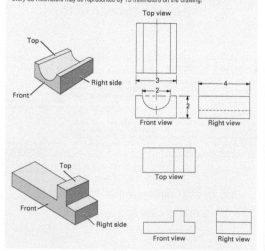

> **Going Green**
> **Residential Solar**
>
> As residential customers look for ways to reduce reliance on fossil fuels like coal and oil, they are turning to solar energy. Usually, they want to install photovoltaic (PV) panels on the roofs of their homes, or in their yards. A residential PV system usually includes the PV panels, a mounting structure that can support the panels, an inverter to convert the direct current (DC) electricity generated by solar photovoltaic modules into alternating current (AC) electricity, and a battery (or batteries) for storing the energy. Most states require a licensed electrician for solar energy system installation, but you should check the local requirements. (Source: **https://www.energy.gov/eere/solar/how-does-solar-work**)

Going Green looks at ways to preserve the environment, save energy, and make good choices regarding the health of the planet.

Cornerstone of Craftsmanship

John Lupacchino
Senior Design Engineer
Gaylor Electric, Inc.

How did you choose a career in the industry?
I knew that this was what I wanted to do since I was in 7th grade.

Who inspired you to enter the industry?
My grandfather, who was an electrician, inspired me to also become an electrician.

What types of training have you been through?
I went to vocational technical school for high school and studied electrical. Then I went on to complete an apprenticeship through the state of Connecticut.

How important is education and training in construction?
Training and education are extremely important in the construction industry. All craftworkers need to participate in continuing education in order to stay up with the advancements in technology.

How important are NCCER credentials to your career?
NCCER credentials are very important because they allow you to showcase your skills and abilities in a standardized way.

What kinds of work have you done in your career?
I have worked in all areas of electrical—residential, commercial, and industrial. Some of the types of facilities I have worked in include houses, stores, factories, warehouses, hospitals, prisons, steel mills, solar installations, cement plants, and many others.

Tell us about your current job.
In my present job as senior design engineer, I am responsible for estimating, designing, and managing electrical construction projects. I am also responsible for electrical code interpretation and compliance.

What do you enjoy most about your job?
My job is challenging and ever changing. It never gets boring. For me, it is great to be a part of building something that benefits others.

What factors have contributed most to your success?
I take advantage of the opportunities that come up, especially the training that is available. Applying my skills and putting in the effort required has definitely contributed to my success.

Would you suggest construction as a career to others? Why?
Yes! The construction industry has limitless opportunities. There will always be a need for building and maintaining facilities, which means there will always be a need for craftworkers.

What advice would you give to those new to the field?
Take advantage of any opportunities for training that you have. Show up to work on time with a "Get It Done" mentality. Do all you can to be the best you can be.

What is an interesting career-related fact or accomplishment?
I have been able to acquire licenses all over the country to enable my employer to work in various locations. I have also had the opportunity to be an instructor in the local ABC apprenticeship program for over 20 years.

How do you define craftsmanship?
Craftsmanship is the quality that comes from creating with passion, care, and attention to detail.

Cornerstone of Craftsmanship boxes feature career stories from people working in related fields.

Know the Code

Electrical Continuity of Metal Raceways, Cable Armor, and Enclosures
NEC Section 300.10

Know the Code boxes feature references to the *National Electrical Code®*.

Review questions at the end of each section and module allow you to measure your progress.

2.0.0 Section Review

1. A straight pull contains two raceways. One of the raceways has a trade size of 3" and one has a trade size of 2". The length of the box *must* be _____.
   a. 16"
   b. 24"
   c. 26"
   d. 32"
2. Where angle or U pulls are made, to determine the distance between raceway entries enclosing the same conductor, you need to multiply the trade size of the largest raceway by _____.
   a. two
   b. four
   c. six
   d. eight
3. If possible, pull boxes should be installed _____.
   a. high on the wall for security
   b. at a height/location that makes it easy to pull conductors
   c. behind wall coverings for a neater look
   d. with extra knockouts left open for air flow

Section Review questions can be found at the end of each section to test your knowledge of the content. Review Questions at the end of each module are provided to reinforce the knowledge you have gained.

Module 26207-23 Review Questions

1. In which *NEC®* article is cable tray installation primarily addressed?
   a. *NEC Article 300*
   b. *NEC Article 392*
   c. *NEC Article 517*
   d. *NEC Article 550*
2. Which type of tray is *best* suited for corrosive areas and atmospheres?
   a. Stainless steel cable tray
   b. Coated aluminum cable tray
   c. Nonmetallic cable tray
   d. Basket tray
3. When a cable tray has a solid bottom, it is referred to as _____.
   a. basket tray
   b. ladder tray
   c. trough
   d. raceway
4. Cable tray is generally manufactured in _____.
   a. 2' and 4' lengths
   b. 6' and 8' lengths
   c. 6' and 12' lengths
   d. 12' and 24' lengths

Additional information can be found in Appendixes at the back of the book.

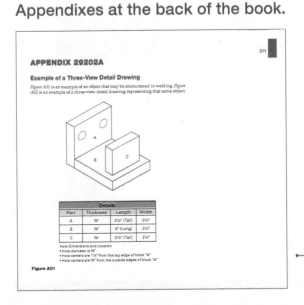

371

APPENDIX 29202A

Example of a Three-View Detail Drawing

Figure A01 is an example of an object that may be encountered in welding. *Figure A02* is an example of a three-view detail drawing representing that same object.

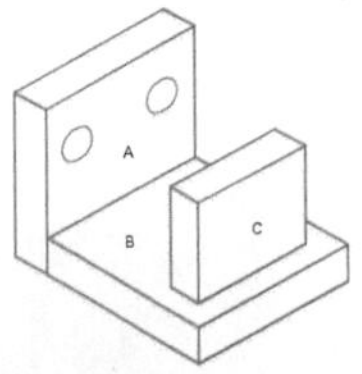

Details			
Part	Thickness	Length	Width
A	¾"	3½" (Tall)	3½"
B	¾"	4" (Long)	3½"
C	¾"	2½" (Tall)	2½"

Hole Dimensions and Location
- Hole diameter is ¾"
- Hole centers are 1¼" from the top edge of block "A"
- Hole centers are ¾" from the outside edges of block "A"

Figure A01

Some modules have corresponding Appendixes that provide supplementary information or activities to enhance your understanding of the material.

NCCERCONNECT

This interactive online course is a unique web-based supplement that provides a range of visual, auditory, and interactive elements to enhance training. Also included is a full eText.

Visit **www.nccerconnect.com** for more information!

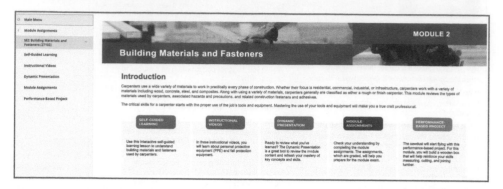

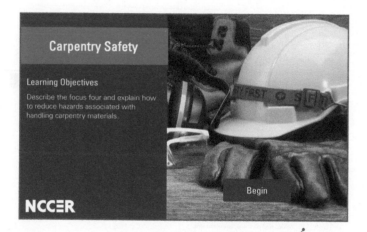

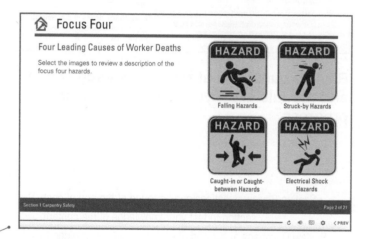

Use the interactive self-guided learning lesson to understand key concepts and terms needed for a career in construction.

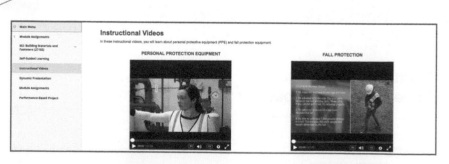

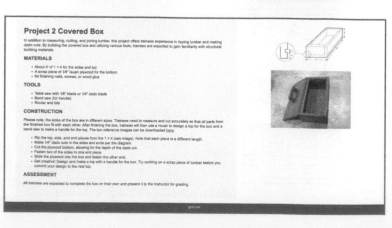

ACKNOWLEDGMENTS

This curriculum was revised as a result of the vision and leadership of the following sponsors:

ABC Carolinas
ABC of Iowa
Cianbro Corporation
CTS Construction
Elm Electrical, Inc.
Gaylor Electric, Inc.
Gould Construction Institute
Intertek

Madison Comprehensive High School
M. Davis & Sons, Inc.
IPS PowerServe
Panduit
Starr Electric Company, Inc.
TIC – The Industrial Company
Tri-City Electrical Contractors, Inc.

This curriculum would not exist were it not for the dedication and unselfish energy of those volunteers who served on the Authoring Team. A sincere thanks is extended to the following:

Paul Asselin
Tim Dean
Kermit Feser
Justin Johnson
Mark Kozloski
Martin Kronz
John Lupacchino II
Tiffany McMillan

Scott Mitchell
John Mueller
Steve Newton
Christine Porter
Karl J Segner
Wayne Stratton
Kevin Tice

A final note: This book is the result of a collaborative effort involving the production, editorial, and development staff at Pearson Education, Inc., and NCCER. Thanks to all of the dedicated people involved in the many stages of this project.

NCCER PARTNERS

To see a full list of NCCER Partners, please visit:

www.nccer.org/about-us/partners.

 You can also scan this code using the camera on your phone or mobile device to view these partnering organizations.

CONTENTS

Contents

Managing Electrical Hazards Appendixes

Review Question Answer Keys 73
Additional Resources 75
Glossary 77

Managing Electrical Hazards

MODULE 26501-24

Source: Oberon

NOTE

NFPA 70E® and *Standard for Electrical Safety in the Workplace*® are registered trademarks of the National Fire Protection Association, Quincy, MA.

Objectives

Successful completion of this module prepares you to do the following:

1. Identify the types and sources of electrical hazards.
   a. Describe shock and arc blast hazards.
   b. Identify and describe equipment-related electrical hazards.
   c. Explain the hierarchy of risk controls.
2. Describe the requirements of NFPA 70E®.
   a. Explain how to identify hazard boundaries.
   b. Describe employer and employee responsibilities with respect to arc flash protection.
   c. Identify the role of human performance as it relates to workplace electrical safety.
3. Identify the causes of electrical incidents and explain how they can be prevented.
   a. Identify the causes of electrical incidents.
   b. Identify safety-related work practices.
   c. Describe the personal protective equipment (PPE) used to protect against electrical hazards.
   d. Describe the other tools and protective equipment used to protect against electrical hazards.
4. Explain the procedures for analyzing electrical hazards.
   a. Identify the steps in a shock risk assessment.
   b. Identify the steps in an arc flash risk assessment.
   c. Complete an Energized Electrical Work Permit.
5. Explain how to establish electrically safe working conditions.
   a. Identify meters used to perform electrical testing.
   b. Explain lockout/tagout (LOTO) procedures.
   c. Describe emergency response procedures and personal safety requirements.

Performance Task

Under supervision, you should be able to do the following:

1. Given a specific electrical task and circumstances, complete an Energized Electrical Work Permit request.

Digital Resources for Managing Electrical Hazards

Scan this code using the camera on your phone or mobile device to view the digital resources related to this craft.

Overview

Working around electrical power presents shock, arc blast, and arc flash hazards. These hazards must be identified and eliminated wherever possible. Where it is not possible to eliminate a hazard, workers must protect themselves using the appropriate safety procedures and personal protective equipment (PPE). NFPA 70E®, *Standard for Electrical Safety in the Workplace*®, provides practical, safe working requirements relating to the use of electricity.

NCCER Industry-Recognized Credentials

If you are training through an NCCER-accredited sponsor, you may be eligible for credentials from NCCER. The ID number for this module is 26501-24. Note that this module may have been used in other NCCER curricula and may apply to other level completions. Contact NCCER at 1.888.622.3720 or go to **www.nccer.org** for more information.

You can also show off your industry-recognized credentials online with NCCER's digital credentials. Transform your knowledge, skills, and achievements into credentials that you can share across social media platforms, send to your network, and add to your resume. For more information, visit **www.nccer.org**.

1.0.0 Types and Sources of Electrical Hazards

Performance Tasks

There are no Performance Tasks in this section.

Objective

Identify the types and sources of electrical hazards.
a. Describe shock and arc blast hazards.
b. Identify and describe equipment-related electrical hazards.
c. Explain the hierarchy of risk controls.

Electricity is generated at a power station and transmitted through power distribution lines to residential, commercial, and industrial users (*Figure 1*). At each point on the power distribution route, maintenance personnel must install, connect/disconnect, adjust, and maintain the equipment used to generate, transmit, and consume electricity. Each point along this route presents hazards that can injure or kill personnel, whether or not they work directly on the equipment. It is essential that these hazards be identified to control the risks associated with working on or near this equipment.

1.1.0 Electrical Shock, Arc Flash, and Arc Blast Hazards

Anyone working on or near electrical equipment may encounter electrical shock, **arc flash**, and **arc blast** hazards. Each type of incident can cause specific types of damage. All workers must be aware of the effects of electrical shock and arc blast hazards in order to provide an effective emergency response.

1.1.1 Electrical Shock

An electrical circuit consists of an electrical power source tied through a designated path to a point of utilization. As long as the current travels the intended path, the circuit is safe. When a person in contact with ground (or grounded objects) contacts an energized part, that person receives an **electrical shock**. Electrical shock can also occur when an internal fault causes the cases or enclosures of tools, appliances, or electrical equipment to become energized. The danger from electrical shock depends on the amount of current flowing through the person, which is a function of voltage and body resistance and may be calculated using Ohm's law. *Figure 2* shows the resistance of the human body to electrical voltage and current. *Table 1* shows the effects of current on the human body.

Arc flash: A dangerous condition caused by the enormous release of thermal energy in an electric arc, usually associated with electrical distribution equipment.

Arc blast: The explosive expansion of air and metal in an arc path. Arc blasts are characterized by the release of a high-pressure wave accompanied by shrapnel, molten metal, and deafening sound levels.

Electrical shock: Occurs when a person or object is grounded and contacts another energized object. The sensation of being shocked occurs when current flows through tissues in the body.

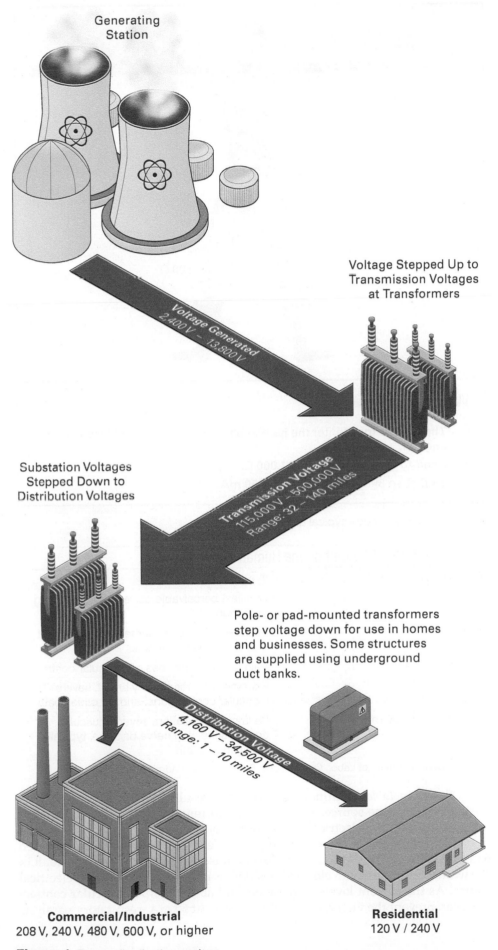

Generating
Station

Voltage Stepped Up to
Transmission Voltages
at Transformers

Voltage Generated
2,400 V – 13,800 V

Substation Voltages
Stepped Down to
Distribution Voltages

Transmission Voltage
115,000 V – 500,000 V
Range: 32 – 140 miles

Pole- or pad-mounted transformers
step voltage down for use in homes
and businesses. Some structures
are supplied using underground
duct banks.

Distribution Voltage
4,160 V – 34,500 V
Range: 1 – 10 miles

Commercial/Industrial
208 V, 240 V, 480 V, 600 V, or higher

Residential
120 V / 240 V

Figure 1 Power distribution system.

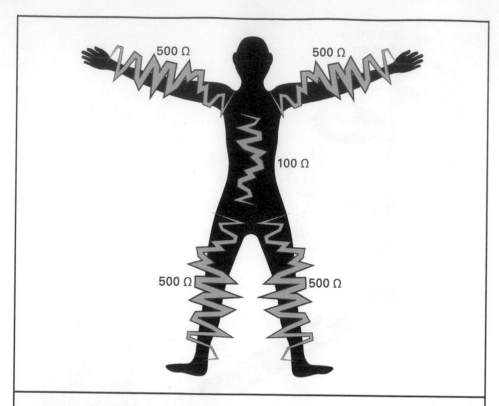

What to Know:

- The current must enter the human body at one place and leave at another.
- Hand-to-hand resistance is 1,000 Ω.
- $I = E/R$, so 50/1,000 = 0.050 A or 50 mA.

Figure 2 Human body's typical electrical resistances.

TABLE 1 Effects of Current on the Human Body

Current Value	Typical Effects
1 mA	Smallest perceivable current. Slight tingling sensation
5 mA	Slight shock and involuntary reactions that may result in other injuries
6 mA to 30 mA	Painful shock and loss of muscular control
50 mA to 150 mA	Extreme pain, respiratory arrest, severe muscular contractions, and possible death
1,000 mA to 4,300 mA	Ventricular fibrillation, severe muscular contractions, and nerve damage, typically resulting in death

Source: US Department of Labor, Occupational Safety & Health Administration (OSHA)

Did You Know?

Electrical Injuries and Fatalities

Electrical injuries are a serious issue for electrical workers. Based on information from the Bureau of Labor Statistics (BLS), electrical injuries result in death more than any of the other injury categories. Additionally, approximately 98% of electrical fatalities happen because of electrical shock, and 70% of those fatalities occur while electrical workers are performing construction, repairing, or cleaning activities.

Source: *NFPA 70E, Informative Annex K*

As shown in *Table 1*, a minor shock of 5 mA results in an involuntary movement away from the source. This can result in injuries as the shocked worker jumps back from the electrical shock source, only to rip open or break an arm or hand on the way out of a cabinet or work area. When the current is between 6 mA and 30 mA, the shock causes a loss of muscular control. This may result in the worker falling from an elevated position or into a more dangerous electrical source. As the current levels increase beyond about 20 mA, muscular contractions can prevent the victim from pulling away. At 50 mA, respiratory paralysis may result in suffocation. Current levels above 1 A may cause the heart to go into a state of fibrillation (abnormal rhythm). This condition is fatal unless the heart rhythm is corrected using a defibrillator. Current levels of 4 A or more may stop the heart, resulting in death unless immediate medical attention is provided.

Other effects of electrical shock include entry and exit wounds from high-voltage contact and thermal burns from current flows of a few amps and up. Thermal burns are often not apparent at first; however, the tissue in the current path may be destroyed and necrotize (rot away) from the inside over time.

1.1.2 Arc Flash and Blast Hazards

Electrical faults may occur when safe work practices are not followed, equipment failure occurs, or both. Two common fault types are **bolted fault** and **arc fault**. A bolted fault occurs when a low-resistance connection (short circuit) is made between two conductors at different potentials. For example, if the conductors of two electrical power phases are crossed during initial installation or repair, the result is a bolted fault. The current flowing between bolted components is called *bolted fault current*. Electrical equipment installed per applicable codes and in good condition should withstand a bolted fault within its rating without damage. Properly sized and applied circuit protective devices are expected to interrupt a bolted fault before it causes damage to equipment or conductors. The high current of a bolted fault enables fast response by these devices.

An arc fault occurs when an electrical current arcs between two energized sources with different potentials or between an energized electrical circuit and ground. The current is called *arc fault current*.

Electric arcs occur in the normal operation of nearly all electrical equipment in situations ranging from plugging a cord into an electrical outlet to operating a switch. This also occurs when meter leads make contact with energized parts. Any time contact is made or broken between energized components and non-energized components, it creates a small arc/spark. Whenever that small arc is not interrupted and grows into a big arc, it creates an event called an *arc flash*.

The duration of the arc and the amount of arc current flow determine the intensity of the arc. If the current level is low and the arc time is short, minimal damage will occur. Any increase in time or current results in greater energy release and destruction and may cause a fire or explosion. An arc flash generates extremely bright light, intense ultraviolet radiation, and temperatures up to 35,000°F (19,400°C). These temperatures melt most metals and all plastics. Blast pressure from an arc may be several hundred pounds per square foot. The blast pressure can blow enclosures apart and people away from the source.

When an arc fault occurs in an enclosed area, the resulting chain of events can cause fire, explosion, toxic fumes, intense light, and shock waves that damage or destroy anything based on the distance from the center of the arc flash (*Figure 3*). The current within an arc fault is less than that in a bolted fault, but the results can be much more destructive. In low-voltage situations, circuit protective devices often take longer to respond to an arc fault than a bolted fault. Longer fault durations increase the risk of personnel injury and equipment damage.

The effects of arc faults must be evaluated to determine the type of personal protective equipment (PPE) required for exposure to an arc hazard. All equipment in which arc faults can occur must be labeled with warnings indicating the arc hazard, anticipated energy exposure, and/or required category of PPE. According to *NFPA 70E Section 130.5(H)*, this type of equipment includes switchboards, panelboards, industrial control panels, meter socket enclosures, and motor control centers (MCC) that are not in dwelling units and are likely to be worked on while energized. The labeling must be large enough to be readable from outside the hazard area. Anyone working on or near the equipment must be trained to recognize and avoid hazards.

Bolted fault: A short-circuit or electrical contact between two conductors at different potentials in which the impedance or resistance between the conductors is essentially zero.

Arc fault: A high-energy discharge between two or more energized conductors or an energized conductor and ground.

Know the Code

Equipment Labeling
NFPA 70E Section 130.5(H)

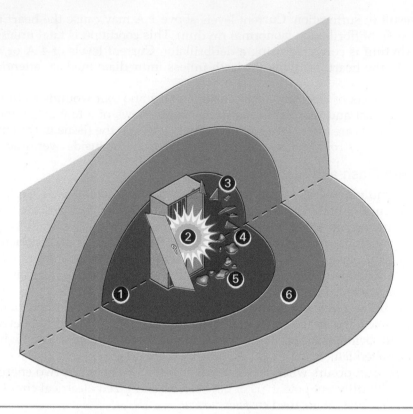

1 **Sound and Pressure Waves**
cause hearing loss and
can rupture ear drums.

2 **Intense Light**
can reach 35,000°F and
cause arc flash burns.

3 **Shrapnel**
impales, lacerates, and
causes thermal contact burns.

4 **Vaporized Copper**
can be inhaled, burning the
mouth, throat, and lungs.

5 **Molten Metal**
sticks to skin and clothing,
causing thermal contact burns.

6 **Hot Air**
rapidly expands, blowing
everything outwards.

Figure 3 Arc flash.

1.2.0 Equipment-Related Electrical Hazards

Electricity is generated at a power station and increased to transmission grid voltage levels (up to 500 kV or more) using step-up transformers. High transmission voltages minimize line loss due to resistance and optimize line capacity. Transmission lines carry voltage to substations, where it is stepped down to local distribution levels. Additional transformers are used to step the voltage down from distribution levels to meet the service needs of customers. Some facilities accept power at the service disconnect, while others use switchgear to accept and distribute power within the facility. This power may then be routed to step-down transformers, lower-voltage switchgear, or MCCs. Power is then routed to panelboards, busways, and cable trays, supplying power to final utilization equipment. Each point along this route presents specific safety concerns, many of which are covered in the following sections.

Lines from Grid

Distribution Lines
to Customer

Power Cables in
Customer's Rack

Switching
Tower

Liquid-Filled Three-Phase
Power Transformer

Distribution Bus

Warning Sign

Figure 4 Electrical substation.

1.2.1 Substations

At switchyards and substations (*Figure 4*), transmission lines connect to the electrical buses, switching equipment, and transformers needed to reduce voltage to local distribution levels.

Larger facilities often have dedicated substations. *Figure 5* shows typical substation equipment for a large facility. Substations are typically located outside but may also be enclosed inside buildings. They are secured to protect the equipment, but more importantly, to keep unqualified personnel out.

Figure 5 Typical substation equipment.
Source: iStock.com/deemac1

Step potential: The voltage between the feet of a person standing near an energized, grounded object, equal to the difference in voltages between each foot and the electrode.

Touch potential: The voltage between the energized object being touched and the feet of the person in contact with it, equal to the difference in voltage between the object and the grounding point.

In a substation, the normally energized high-voltage parts are exposed and easily identified. Clearances between energized parts and ground are maintained by the structure design to ensure that workers walking past do not approach nearer than a safe distance. The substation has a fence to keep people out and a grounding grid to dissipate voltage surges due to switching or a lightning strike. The grounding grid also bonds all conductive noncurrent-carrying equipment and structure(s) to earth and the source. The grounding and bonding systems in substations must carry electric current into earth under both normal and fault conditions. A broken or disconnected grounding/bonding connection can be a significant hazard to a worker bridging the path to ground under fault conditions or when induced voltage is possible. It can also result in loss of service and equipment damage.

The ground grid extends several feet outside of the fence perimeter and is bonded to the fence at frequent intervals. The intent is to eliminate both **step potential** and **touch potential** for anyone approaching grounded parts during fault conditions. Step potential is the voltage between the feet of a person standing near an energized, grounded object. It is equal to the difference in voltage between two points at different distances from the electrode. The substation yard is covered with several inches of crushed gravel to provide a level of insulation or high resistance between a person and earth. To help protect against step potential, dielectric footwear is typically required in substations and switchyards.

Touch potential is the voltage between the energized object being touched and the feet of the person in contact with it. It is equal to the difference in voltage between the object and the grounding point. The touch potential can be very high when the object is grounded at a point remote from where the person is in contact with it.

The high voltage levels and currents associated with overhead lines and substations present some unique hazards. In a substation, it is easy to identify normally energized parts and to visually identify de-energized circuit parts by observing open switches. However, de-energized circuit parts can become energized due to static charge, lightning strikes, or electromagnetic fields.

Static charges can be induced in ungrounded conductors due to wind. Static charges developed along a length of de-energized overhead line can cause a fatal shock should a worker become the path from line to ground.

Electrical field charges develop directly proportional to line voltage and proximity to the line, and may accumulate on ungrounded conductive material and shock a grounded person coming in contact with them.

Design standards govern minimum heights of transmission lines and maximum levels of field strength under lines and along rights-of-way.

In addition, magnetic fields can cause induced voltages in conductors and result in circulating currents in grounding systems. The level of induced voltage is low in grounded conductors or structures and becomes high when the ground path is broken.

Temporary protective grounds are applied to de-energized overhead circuit parts that may be approached nearer than a safe distance during installation, repair, or maintenance. The purpose of temporary grounds is to facilitate the operation of protective devices in the event of inadvertent reenergization and to dissipate electromagnetic charges and lightning strikes (either direct or nearby).

1.2.2 Service-Entrance Disconnects

Electrical power from distribution lines into a facility is routed through a service-entrance disconnect. Service disconnects may be installed outside or inside. Local codes specify their location. The disconnect handle may or may not be locked. Service disconnects are rarely opened unless personnel must perform maintenance on downstream equipment. *Figure 6* shows service disconnects mounted outside on a concrete pad.

Window Allows Visual Verification That Switch Is Open

Interior Door of Load Break Disconnect Switch

Figure 6 Pad-mounted service disconnects.
Source: Jim Mitchem

As with any switching device, service disconnects are subject to faults or component failures that could cause a fire or explosion. For those reasons, the hazard boundaries around service disconnects must be identified and warning labels posted (*Figure 7*).

WARNING!

When an arc fault occurs upstream of the main service disconnect, the utility company may not have devices that clear the fault quickly, leading to very high arc energies. Facility electricians have no access to or control of utility protective devices.

If work must be performed on a service disconnect, the power source must be shut down and locked out until after the work is completed. Lockout of utility-supplied power may need to be performed by utility personnel.

1.2.3 Transformers

Electrical power being fed into a facility from a substation or a service disconnect may need to be stepped down before it is applied to the process equipment. The transformers used for industrial or commercial customers range from small pad- or pole-mounted transformers to very large transformers found in substations or in the power rooms of large facilities. The areas surrounding the transformers must be kept clear of tools, trash, and other equipment. *Figure 8* shows a transformer located within an office building.

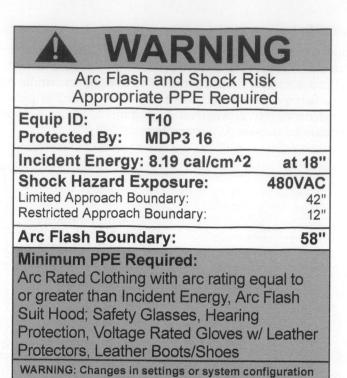

<table>
<tr><td colspan="3" align="center">⚠ WARNING</td></tr>
<tr><td colspan="3" align="center">Arc Flash and Shock Risk
Appropriate PPE Required</td></tr>
</table>

Equip ID:	**T10**	
Protected By:	**MDP3 16**	
Incident Energy: 8.19 cal/cm^2		**at 18"**
Shock Hazard Exposure:		**480VAC**
Limited Approach Boundary:		42"
Restricted Approach Boundary:		12"
Arc Flash Boundary:		**58"**

Minimum PPE Required:
Arc Rated Clothing with arc rating equal to or greater than Incident Energy, Arc Flash Suit Hood; Safety Glasses, Hearing Protection, Voltage Rated Gloves w/ Leather Protectors, Leather Boots/Shoes

WARNING: Changes in settings or system configuration may invalidate the results.   Date: 11/10/24   #0002

Figure 7 Service equipment warning label.
Source: John Lupacchino

Figure 8 Transformer.

Transformers are reliable and present little risk of **shock hazard** because they either do not have exposed energized parts or their exposed energized parts are enclosed within a vault or other protected area.

A transformer may fail due to inadequate maintenance or failure of protective devices. Some transformers contain hazardous compounds. If a transformer has burned, treat the soot and burned material as potentially hazardous until proven otherwise.

Shock hazard: A dangerous condition associated with the possible release of energy caused by contact or approach to energized electrical conductors or circuit parts.

1.2.4 Switchgear

Switchgear is used to accept and distribute electrical power in larger facilities. Metal-clad switchgear may be found in voltage ratings from 1,000 V to over 30 kV. The service voltage is distributed through switches and circuit breakers in switchgear to other distribution equipment, such as step-down transformers, lower-voltage switchgear, and MCCs.

Switchgear contains protective relaying devices and instrument transformers. *Figure 9* shows a typical switchgear enclosure similar to those found in many industrial facilities.

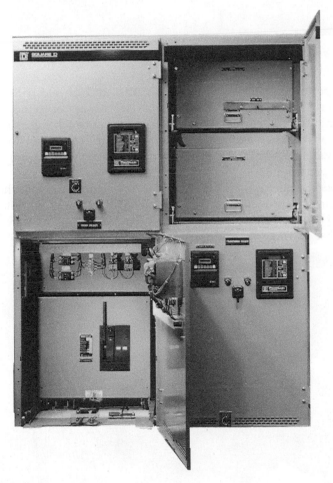

Figure 9 Switchgear enclosure.
Source: Schneider Electric

Switchgear typically contains high levels of current and presents a significant **arc flash hazard**. Warning labels are required on the front of the switchgear. As long as there are no energized parts exposed on the operating side (exterior) of the switchgear, it is considered to have a **dead front**, meaning it is free from an electrical shock hazard while the cover(s) are closed.

Any unused opening into an enclosure must be sealed to meet the environmental requirements of the enclosure. Sealing the enclosure also prevents the entrance of rodents or other creatures that can chew insulation or contact exposed energized parts, resulting in faults that destroy equipment and cause downtime (*Figure 10*).

Arc flash hazard: A dangerous condition associated with the release of energy caused by an electric arc.

Dead front: Equipment that has no exposed energized electrical conductors or circuit parts on the operating side.

Figure 10 Equipment damage caused by rodent entry.
Source: Tri-City Electrical Contractors Inc.

During periodic maintenance, electricians and/or electrical technicians inspect, clean, tighten, lubricate, and possibly adjust or repair components inside the enclosure. Terminal fasteners used to secure cabling must be checked annually to ensure they are tight. A loose connector can cause an arc. Infrared imaging devices (*Figure 11*) are often used to identify overheating due to loose fasteners or corrosion.

As long as the enclosure is closed and properly maintained, the hazards of switchgear are minimized. Many of the arc flash incidents in switchgear occur when circuit breakers are racked in or out. *Figure 12* shows a low-voltage power circuit breaker carriage in the drawout position.

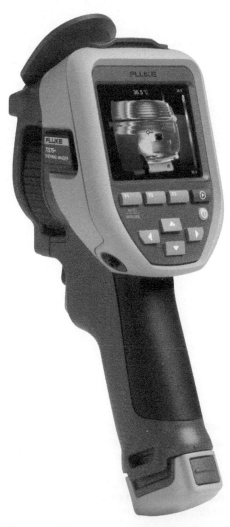

Figure 11 Infrared imaging device.
Source: Reproduced with Permission, Fluke Corporation

Figure 12 Circuit breaker carriage.

1.2.5 Motor Control Centers

MCCs are commonly found in both 600 V and 5 kV ratings. Medium-voltage MCCs are similar in construction to medium-voltage switchgear and present similar hazards.

MCCs supply power to operate electric motors and other process machinery. When a motor or process equipment must be locked out and secured, the lockout point is usually at the MCC cubicle or bucket supplying the equipment. Most safety policies require that the equipment receiving power from an MCC be labeled as to which MCC is feeding it. Individual buckets in the MCC must also be labeled to indicate the equipment they serve. *Figure 13* shows a typical low-voltage (600 V and below) MCC.

Three-Phase Supply from MCC Bus

Disconnect Switch for MCC Bucket

(A)   **(B)**

Figure 13 MCC.
Source: Revere Control Systems (13A)

Low-voltage MCCs present significant electrical hazards. Individual buckets may contain both power voltage and control voltage. Sometimes, control voltage is supplied from a source other than a control transformer within the bucket. When the bucket contains voltage sources that are not removed by opening the disconnect device, warning labels must indicate that the equipment is fed from multiple sources.

It is common to find MCCs with available fault currents of 40,000 A (40 kA) and more. This energy presents a significant arc flash potential.

WARNING!

When the main bus is energized in switchgear or an MCC, every compartment and open door is considered an arc hazard.

MCCs may be fed directly from power transformers, in which case the MCC has a main breaker operating as both the service disconnect and the protective device for the MCC bus. In many cases, the main breaker is set to respond instantaneously to both bolted and arcing faults on the bus, thus limiting equipment damage and flash energy exposure. In other cases, the MCC does not have a main breaker and is supplied from a feeder breaker in low-voltage switchgear. These breakers commonly have a time delay of up to a $\frac{1}{2}$ second before the feeder breaker operates to clear a fault. An uninterrupted arcing fault may continue for some time, increasing damage and arc flash energy exposure.

MCC enclosures house multiple buckets serving motors and other equipment. Opening the disconnect switch of an individual bucket removes the electrical hazard from only the equipment it serves. As long as the MCC bus is energized, there is an arc flash hazard anywhere within that MCC. There may or may not be a shock hazard in an individual bucket when all sources of supply are disconnected from that bucket. The duration of any arc fault within an MCC is determined by the protective device supplying the MCC bus. When an arc fault occurs within a MCC or other equipment, the arc escapes wherever possible, breaching doors and covers to do so. *Figure 14* shows a fault that occurred between the bus and bucket.

WARNING!

Removing or inserting buckets when a bus is energized presents a significant arc flash hazard and is specifically warned against by most MCC manufacturers.

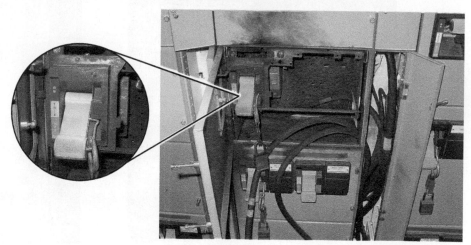

Figure 14 Effects of fault between bus and bucket.
Source: Jim Mitchem

After electrical power from the MCC to the equipment has been turned off and locked out, always verify that the electrical power from the MCC is removed from the equipment.

When operating any electrical disconnect switch, stand on the handle side so no part of your body crosses in front of the equipment. This reduces the risk of being in the line of any arc flash or explosion that could occur when the switch components are opened or closed.

1.2.6 Cable Tray

Cable tray is often used to route power in commercial and industrial facilities. *Figure 15* shows a typical cable tray system. Cable tray should be labeled by name and voltage class. Clear identification of all components is a major factor in effective electrical safety.

Periodically inspect cable tray and cabling systems for signs of damage or aging of insulation. Poorly supported or rusted cable tray can collapse and damage conductors and equipment. If cables age to the point where the insulation weakens or becomes cracked or cut, the cables should be replaced.

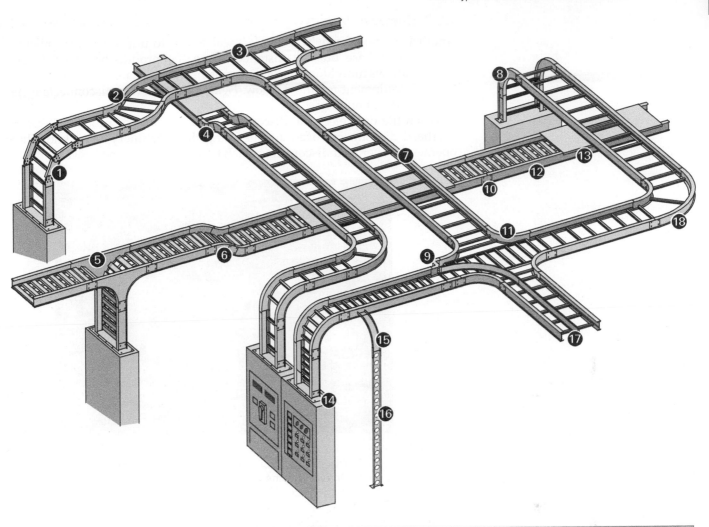

1	Vertical Bend Segment (VBS)		**10**	Straight Splice Plate
2	45° Horizontal Bend, Ladder Cable Tray		**11**	Horizontal Cross, Ladder Cable Tray
3	Horizontal Tee, Ladder Cable Tray		**12**	Ventilated Trough Cable
4	30° Vertical Inside Bend, Ladder Cable Tray		**13**	Solid Flanged Tray Cover
5	Vertical Tee Down, Ventilated Trough Cable Tray		**14**	Frame-Type Box Connector
6	45° Vertical Outside Bend, Ventilated Cable Tray		**15**	Channel Cable Tray, 90° Vertical Outside Bend
7	Ladder Cable Tray		**16**	Ventilated Channel Straight Section
8	90° Vertical Outside Bend, Ladder Cable Tray		**17**	Barrier Strip Straight Section
9	Left-Hand Reducer, Ladder Cable Tray		**18**	90° Horizontal Bend, Ladder Cable Tray

Figure 15 Cable tray system.

1.2.7 Busway

Another method for delivering electrical power to utilization equipment is to use a three-phase busway installed above the production areas. Overhead busway allows individual disconnect switches to be located wherever necessary to serve equipment mounted below. *Figure 16* shows two disconnects installed on an overhead busway.

When the only electrical disconnect for a machine or process is one of the overhead disconnect boxes, anyone using the machinery must know how to operate the overhead disconnect switch to remove incoming power.

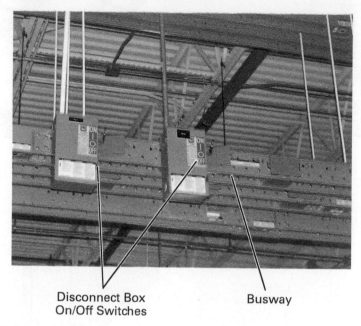

Disconnect Box
On/Off Switches   Busway

Figure 16 Busway.

1.2.8 Circuit Breakers and Panelboards

Panelboards contain busbars, circuit breakers, and conductors. Panel trim and dead-front covers eliminate shock hazard until covers or trim are removed. In existing facilities, it is common for a new circuit to be added or a failed circuit breaker to be replaced. An electrical shock hazard is present any time the covers are removed while the panel is energized. The response time of upstream circuit breakers or other protective devices must be considered when evaluating panelboards for arc flash hazards. *Figure 17* shows breaker enclosures with trim and covers removed.

1.2.9 Electrical Terminal Cabinets and Junction Boxes

The power fed from MCCs or panelboards to utilization equipment may pass through terminal cabinets or junction boxes (*Figure 18*) that house the control devices for the motors or process machinery. When accessing terminal cabinets and junction boxes to test, adjust, troubleshoot, or replace control devices, you may be exposed to the power passing through the enclosure. In some facilities, nonelectrical technicians work within inches of exposed energized parts or arc hazards. While those technicians work near but not on energized parts, they are still exposed to electrical hazards and must be qualified, aware of the hazard, and use appropriate PPE and safe work practices.

Breakers without Outer Cover

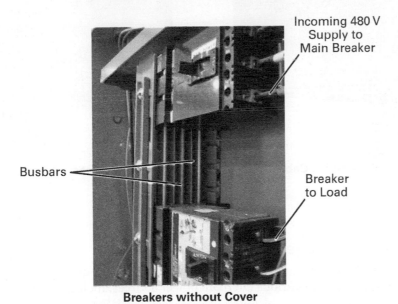

Incoming 480 V
Supply to
Main Breaker

Busbars

Breaker
to Load

Breakers without Cover

Figure 17 Breaker enclosures with covers removed.

1.2.10 Field-Mounted Safety Switches

Safety (disconnect) switches allow the power to be removed at or near the equipment. *Figure 19* shows typical field-mounted disconnect switches. These switches may or may not contain overload devices.

WARNING!

When opening the cover of a disconnect switch, there is exposure to shock and arc fault hazards. For example, one or more blades of a switch may fail to separate when the operating handle is moved to the open position. Remember, it is not safe to proceed until the disconnect is locked out and the absence of voltage has been verified at the work location.

(A) Junction Box

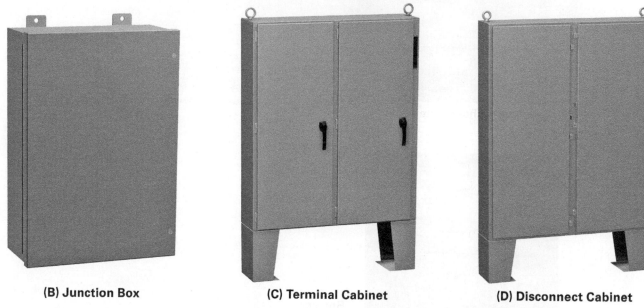

(B) Junction Box   **(C) Terminal Cabinet**   **(D) Disconnect Cabinet**

Figure 18 Terminal cabinets and junction boxes.
Sources: Jim Mitchem (18A); Hammond Manufacturing Company, Inc. (18B–D)

1.2.11 Alternate Power Sources

In addition to electrical power distribution and control equipment, plant personnel also work with various alternate power sources, including:

- *Backup generators* — In some environments, electrical power to a process or facility must be maintained at all times. When the electricity supplied by a utility fails, a combustion engine-powered generator (*Figure 20*) must start and stabilize immediately to maintain the supplied electrical power. Plant personnel start and test these generators as part of routine preventive maintenance. When backup generators are used, automatic or manual transfer switches (*Figure 21*) are employed to remove the generators' user from the utility grid and switch to generator-supplied power. Generators use combustible fuel and present fire and explosion hazards. In addition, they present hazards from combustion byproducts and require a carefully designed ventilation system. Electrical hazards include overload, shock, and arc flash hazards. All electrical hazards must be identified, and workers must use the appropriate PPE when working with generators and associated switchgear. The arc flash hazard of equipment is often different when operating on generator power than when operating on utility power.

Figure 19 Field-mounted safety switches.

- *Batteries and battery-charging areas* — Storage batteries of various types are often used in industrial plants and other facilities. Batteries are used in uninterruptible power supply (UPS) systems and to operate battery-powered equipment. For example, battery-powered forklifts are often used to keep combustion fumes to a minimum. These batteries must be charged, disconnected, installed in the vehicle or device being used, and then later removed and reconnected to a charger. Specially trained operators often perform these actions, but maintenance personnel may be called in to do similar work or to work on the devices using or recharging the batteries. Special care must be taken when working on or near batteries. Many batteries contain acid, which must be stored and handled carefully to avoid burns. Plates within the batteries can short and cause the battery to explode when connected to the charger cables. An exploding battery can spread acid as it explodes. Fire and electrical codes dictate where and how batteries are exchanged and recharged. Environmental codes dictate acid storage, use, and disposal requirements.

> **NOTE**
>
> Newer battery technologies have different hazards and handling requirements. Always read and follow the safety requirements for the battery in use.

WARNING!

Batteries are designed to produce high current levels and are dangerous because they have no On/Off switch. If something or someone makes contact with the positive and negative terminals, there is no way to turn off the power. The only way to stop the current flow is to break the circuit.

- *Uninterruptible power supply systems* — UPS systems protect equipment from damage by providing backup power during power failure. Process systems using electronic instrumentation and computer control systems must have a reliable power source to operate field instruments, monitor the process, and bring the process to a safe shutdown when electrical power is interrupted. These systems typically have backup power from a UPS system. *Figure 22* shows sealed batteries in a UPS cabinet.

Lead-acid battery banks pose both chemical and shock hazards. In addition, large UPS systems present electrical shock and arc flash hazards from the incoming power supply. These systems are designed to isolate and bypass the charging system to allow for safe maintenance and to reduce the risk of equipment failure due to operator error. When working with batteries and in battery rooms, refer to *NFPA 70E Articles 240 and 320.*

Know the Code

Batteries and Battery Rooms
NFPA 70E Article 240

Know the Code

Safety Requirements Related to Batteries and Battery Rooms
NFPA 70E Article 320

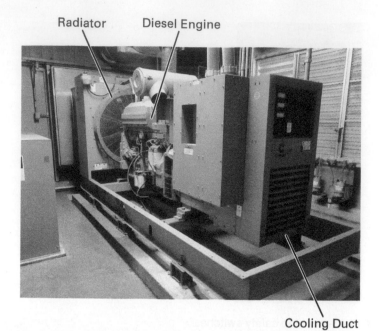

Figure 20 Engine-generator set.

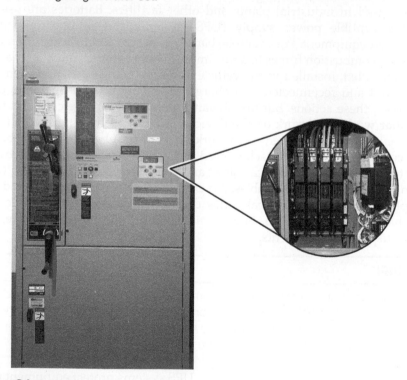

Figure 21 Transfer switch.
Source: Tri-City Electrical Contractors Inc.

Sealed Lead-Acid Batteries

Battery Cabinet　　　　　　　　　　　UPS

Figure 22 UPS batteries.
Source: Tim Ely

1.3.0　Hierarchy of Risk Controls

Following a hierarchy of risk controls leads to the implementation of safer systems and can reduce or eliminate electrical hazards. The major methods of controlling hazards are listed in *NFPA 70E Section 110.3(H)(3)* and *NFPA 70E Table F.3* in the Annex. This hierarchy is illustrated in *Figure 23*. They include the following:

- Elimination
- Substitution
- Engineering controls
- Awareness
- Administrative controls
- PPE

Know the Code

Hierarchy of Risk Control Methods
NFPA 70E Section 110.3(H)(3)

Know the Code

The Hierarchy of Risk Control Methods
NFPA 70E Table F.3

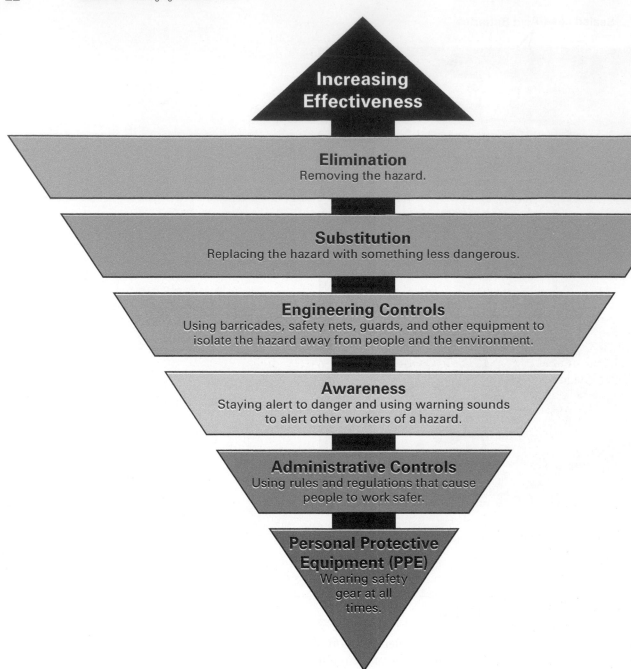

Figure 23 Hierarchy of risk control methods.

According to Informational Note No. 1 in *NFPA 70E Section 110.3(H)(3)*, elimination, substitution, and engineering controls are the most effective methods of reducing risk. These controls are usually applied at the source of the hazard and are, therefore, less likely to be affected by human error. Awareness, administrative controls, and PPE are the least effective methods of reducing risk as they are more likely to be affected by human error.

NFPA 70E Informative Annex F provides additional information on risk assessment and risk control.

1.3.1 Elimination

The best method of reducing a risk is to eliminate the hazard. For example, this can be done by establishing an **electrically safe work condition (ESWC)** using appropriate lockout/tagout (LOTO) procedures. Another example would be to schedule off-shift maintenance to eliminate the hazards associated with working near operating equipment.

1.3.2 Substitution

Substitution is used to replace a hazard with a less hazardous or nonhazardous material, system, or process. Examples include choosing a nontoxic coating over a volatile coating, reducing energy by replacing a 120 V control circuit with 24 V control circuitry, replacing one large storage tank with several smaller cylinders, and using meters with remote reading capability. Another example is the use of robotic systems that allow workers to perform remote racking from a safe distance.

1.3.3 Engineering Controls

Engineering controls include methods of isolating the hazard from the worker. Common types of engineering controls include barricades, safety nets, machine guards, interlocks, and hand switches located outside the danger zone that must be touched and/or held down for the machine to operate. Engineering controls require compliance on the part of workers. When engineering controls are used, the company must develop and enforce policies and procedures mandating their use.

1.3.4 Awareness

Awareness can help to reduce risk by warning workers of various jobsite hazards. Examples include signs, such as electrical hazard warning labels. Signs must be highly visible and easy to understand. Symbols may be used to help any workers with language barriers.

Warning devices are another form of awareness control. These devices include horns, bells, whistles, and lights. In some cases, workers hear alarms or signals so often they unconsciously ignore them. For these warning devices to be effective, they must be distinctive and audible over ambient noise levels.

1.3.5 Administrative Controls

Administrative controls use practices and policies to limit an employee's exposure to a hazard. Safety policies and procedures, operating procedures, and maintenance procedures are examples of administrative controls, which are documented and formalized by management. Further examples include LOTO, confined space entry procedures, and work permits.

Worker rotation is also an administrative control. Workers should be rotated if the work involves hot or cold environments, constant or repetitive motion, or is either very stressful or tedious. This can be accomplished by cross-training workers for different jobs on the site, adjusting the work schedule, or providing frequent breaks.

Administrative controls require training and enforcement to be effective. When administrative controls are specified, they should be audited to verify their effectiveness.

1.3.6 Personal Protective Equipment

PPE is designed to prevent workers from coming into contact with a hazard. Examples include shock and arc flash PPE. PPE protects individuals but does not create a safe work environment. PPE is an important component of risk control but should be considered the last line of defense in reducing hazards.

> **Methods of Reducing Clearing Time**
>
> For any overcurrent device rated or that can be adjusted to 1,200 A or higher (regardless of voltage), *NEC Section 240.87(B)* provides allowable methods of reducing the circuit breaker clearing time when an ESWC cannot be attained and the work must be done energized. This allows the circuit breaker to operate faster should an arc fault occur while work is being performed. Note that this temporarily voids selective coordination and may only be performed under engineering supervision and an approved Energized Electrical Work Permit.

> **Prevention through Design**
>
> The National Institute for Occupational Safety and Health (NIOSH) leads a national initiative known as *Prevention through Design (PtD)*. The goal of the initiative is to prevent or reduce occupational injuries, illnesses, and fatalities by eliminating the hazard at the source. For example, many power tools are very loud and present a hearing hazard that must be managed through the use of hearing protection. The NIOSH Buy Quiet initiative encourages using power tools designed to operate at a much lower noise level.

1.0.0 Section Review

1. Ventricular fibrillation is likely to occur when the body is exposed to current levels of _____.
   a. 1 mA
   b. 5 mA
   c. 500 mA
   d. 1 A

2. Touch potential can be very high when the object is grounded _____.
   a. directly at the point of contact
   b. at a location remote from the point of contact
   c. within the fence perimeter
   d. directly below the point of contact

3. A barricade is an example of _____.
   a. an administrative control
   b. an engineering control
   c. worker awareness
   d. hazard elimination

2.0.0 Getting Started with NFPA 70E®

Performance Tasks

There are no Performance Tasks in this section.

Objective

Describe the requirements of NFPA 70E®.
   a. Explain how to identify hazard boundaries.
   b. Describe employer and employee responsibilities with respect to arc flash protection.
   c. Identify the role of human performance as it relates to workplace electrical safety.

Know the Code

Introduction
NFPA 70E Article 90

Know the Code

Definitions
NFPA 70E Article 100

The Occupational Safety and Health Administration (OSHA) issues regulations addressing safety in the workplace. While OSHA is a federal agency and its regulations are law, OSHA also relies on national consensus standards for certain requirements. For electrical safety, OSHA recognizes several standards from the National Fire Protection Association (NFPA). NFPA 70®, *The National Electrical Code®* (*NEC®*), provides requirements for electrical installations, while NFPA 70E®, *Standard for Electrical Safety in the Workplace®*, provides practical, safe working requirements relative to the hazards arising from the use of electricity.

NFPA 70E® is divided into three chapters:

- *NFPA 70E Chapter 1, Safety-Related Work Practices*
- *NFPA 70E Chapter 2, Safety-Related Maintenance Requirements*
- *NFPA 70E Chapter 3, Safety Requirements for Special Equipment*

Like the *NEC®*, each chapter is further subdivided into articles and sections. There are also numerous annexes that provide calculation methods, forms, and other references. This module focuses on the safety-related work practices covered in *NFPA 70E Chapter 1*.

NFPA 70E Article 90 describes the scope and structure of the standard. *NFPA 70E Article 100* defines terms used throughout the standard. Take the time to review these definitions; many of them are specific to the hazards discussed in this standard.

NFPA 70E Article 105 covers the application of safety-related work practices and procedures. *NFPA 70E Section 105.3* discusses the responsibilities of the employer and employee.

General requirements for employee safety relative to electrical hazards in the workplace are covered in *NFPA 70E Article 110*. *NFPA 70E Section 110.2* emphasizes that elimination of hazards is the first priority. Host employer and contract employer responsibilities are covered in *NFPA 70E Section 110.5*.

NFPA 70E Section 110.4(A) identifies safety training requirements for employees who could encounter electrical hazards not reduced to a safe level through applicable electrical installation requirements. This applies to any employee who may approach nearer than a safe distance or who is expected to test, troubleshoot, or repair electrical equipment. It also applies to operators who may perform switching operations. The degree of training must be appropriate to the hazards and risk encountered by the employee. *NFPA 70E Section 110.4(A)(3)* also identifies the requirements for additional training and retraining. The intent of retraining is to address changes to NFPA 70E® as well as changes in the employer's policies and procedures concerning electrical safety. The interval for retraining must not exceed three years.

NFPA 70E Section 110.4(A) addresses training requirements and characteristics of both qualified and unqualified personnel. It is important to note that there is no value judgment attached to these terms, and they have no relationship to license status or time in the trade. A **qualified person** has demonstrated skills and knowledge related to the construction and operation of electrical equipment and installations and has received safety training to identify the hazards and reduce the associated risk. An **unqualified person** is simply one who is not a qualified person. Unqualified persons must receive training in any safety-related work practices necessary to ensure their safety. Note that a person may be qualified to perform one task on a piece of equipment but may not be qualified to perform another task on the same piece of equipment. Similarly, a person may be qualified to perform a task on one brand of equipment but may not be qualified to perform the identical task on another brand of the same type of equipment.

NFPA 70E Articles 120 and 130 address work involving electrical hazards. These sections identify requirements such as creating an ESWC, performing an electrical risk assessment, and the use of test equipment. The procedures to meet the requirements of *NFPA 70E Article 130* are discussed in the remainder of this module, supplemented by *NFPA 70E Article 120* and the Informative Annexes found in the back of the standard.

2.1.0 Recognizing Hazard Boundaries

Ideally, work on or near electrical equipment would always be performed with no electrical power applied (also known as an *ESWC*), but that is not always possible. NFPA 70E® requires special safety procedures when working on or near circuits with voltage levels of more than 50 V line-to-line.

Distance is the best protection against electrical hazards. NFPA 70E® establishes specific limits of approach to exposed energized parts. These limits are for shock protection and are called *approach boundaries*. The approach boundaries, required PPE, and other tools or equipment are determined by performing a shock risk assessment.

When working with electrical equipment, assume the equipment is energized until personally verifying otherwise. As noted earlier, electrical equipment presents a potential shock, arc flash, and arc blast hazard. *Figure 24* shows a diagram indicating the approach limits defined by NFPA 70E®. Note that the **arc flash boundary (AFB)** is determined independently and may be at a greater or lesser distance than shock protection boundaries.

Know the Code

Application of Safety-Related Work Practices and Procedures
NFPA 70E Article 105

Know the Code

General Requirements for Electrical Safety-Related Work Practices
NFPA 70E Article 110

Qualified person: A person who has the necessary training or certifications and has demonstrated the knowledge and ability to safely install and operate electrical equipment.

Unqualified person: A person who does not possess the skills, knowledge, or training necessary to be a qualified person.

Know the Code

Establishing an Electrically Safe Work Condition
NFPA 70E Article 120

Know the Code

Work Involving Electrical Hazards
NFPA 70E Article 130

Arc flash boundary (AFB): An approach limit at a distance from exposed energized electrical conductors or circuit parts within which a person could receive a second-degree burn if an electrical arc flash were to occur.

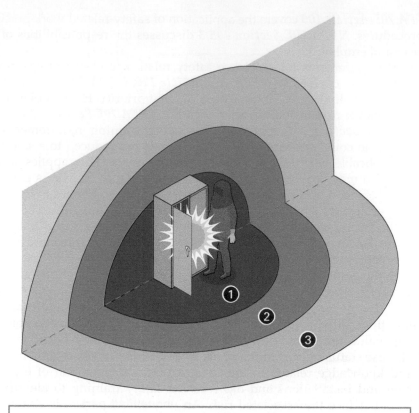

① Restricted Approach Boundary (Qualified Persons with PPE)

② Limited Approach Boundary (Qualified Persons)

③ Arc Flash Boundary (Could Be Closer or Farther, Depending on Incident Energy)

NOTE: Shock boundaries are dependent on system voltage level and refer to distance from exposed energized parts.

Figure 24 Approach limits.

The exposed energized component can be a wire or a mechanical component inside the electrical equipment. All boundary distances are measured from that point. When establishing boundaries, exposed movable conductors are treated differently than exposed fixed circuit parts or conductors.

Every possible electrical hazard within a work area must be analyzed and documented. Specific PPE for a given situation is based on the information gathered from the analysis of a given hazard. That documented data includes all the electrical hazards (arc flash, blast, and shock). After all hazards have been documented, all personnel (qualified and unqualified) working in the area must be trained to recognize and avoid the identified hazards. *NFPA 70E Section 130.4(F)* establishes rules for unqualified persons who must enter the **limited approach boundary**. Only qualified persons using all required PPE are allowed to enter and work inside the **restricted approach boundary** per *NFPA 70E Section 130.4(G)*.

Know the Code

Limited Approach Boundary
NFPA 70E Section 130.4(F)

Know the Code

Restricted Approach Boundary
NFPA 70E Section 130.4(G)

Limited approach boundary: An approach limit at a distance from an exposed energized electrical conductor or circuit part within which an electrical shock hazard exists.

Restricted approach boundary: An approach limit at a distance from an exposed energized electrical conductor or circuit part within which there is an increased likelihood of electric shock.

2.1.1 Arc Flash Boundary

When an arc flash hazard is present, an AFB must be established. This boundary is determined by how far away a person would need to be located to avoid receiving serious burns in the event of an arc flash. Anyone within the AFB is exposed to the possibility of second-degree (blistering) burns or worse and must, therefore, use arc-rated protective apparel as determined through the **arc flash risk assessment** and other PPE identified in *NFPA 70E Article 130*.

Depending on the **incident energy**, an AFB might be within or outside a shock protection boundary. Many electrical safety programs establish both the AFB and the outer shock protection boundary at whatever distance is greater, as determined by the risk assessment.

When an electrical fault causes an arc flash (*Figure 25*), the explosion produces both a fireball and a shock wave extending away from the arc flash location (the conductor or circuit). Anyone within the AFB will be exposed to searing heat and extremely bright light that may cause pain and temporary loss of vision. The heat from arc flashes is often hot enough to melt metal fixtures inside the enclosure.

Arc flash risk assessment: A study investigating a worker's potential exposure to arc flash energy, conducted for the purpose of injury prevention, determination of safe work practices, and appropriate levels of PPE.

Incident energy: The amount of thermal energy impressed on a surface at a certain distance from the source of an electrical arc. Incident energy is typically expressed in cal/cm^2.

Figure 25 Arc flash.
Source: Honeywell | Salisbury

Arc Flash Relays

The energy discharged in an arc is calculated by multiplying the square of the short-circuit current by the time the arc takes to develop (energy = i^2t). Therefore, the energy and, consequently, the power and size of an arc flash are directly related to both short-circuit current and time. The higher the arcing current or the longer the duration of the arc, the more damage is likely to occur. An arc flash relay uses light sensors or a combination of light and current sensors to detect the light/current produced by an arc flash and trip upstream devices. This reduces the incident energy by reducing the clearing time of the overcurrent device. It also limits the potential damage. See *NFPA 70E Informative Annex 0.2.3(6)*.

The AFB is determined for thermal energy, but a blast accompanies the arc flash. The blast creates a shock wave that can blow equipment apart and people away from the blast. Shrapnel, toxic gases, and copper vapor explode in all directions. The blast also creates sound waves that can damage hearing.

2.1.2 Shock Protection Boundaries

There are two electrical shock protection boundaries or limits of approach. NFPA 70E® identifies these electrical shock boundaries as follows:

- Limited approach boundary
- Restricted approach boundary

The limited approach boundary is a shock protection boundary at a specified distance from an exposed energized part that should be crossed only by qualified persons. All unqualified persons in the area must be made aware of the hazards and warned not to cross the boundary. Where there is a need for an unqualified person to cross the limited approach boundary, a qualified person must advise the unqualified person of the possible hazards and continuously escort the unqualified person while inside the limited approach boundary. Although an unqualified person may be escorted past the limited approach boundary (into the shock hazard area), they may *not* be taken beyond the AFB (into the arc flash hazard area). Check all of your approach boundaries before escorting an unqualified person into the limited approach boundary.

WARNING!

Any worker exposed to electrical hazards must be qualified to manage the hazard and use appropriate PPE. If any worker is not exposed to an electrical hazard but has the potential to be, the worker must be informed of the hazard and instructed on how to avoid it.

A restricted approach boundary is a shock protection boundary that, due to its proximity to exposed energized parts, requires the use of shock protection techniques and equipment when crossed. The restricted approach boundary may be crossed only by qualified persons using the required PPE (*Figure 26*).

Work within the restricted approach boundary requires that rubber insulating equipment be used inside that boundary. It also requires the use of insulated tools rated for the voltages on which they are used.

The restricted approach boundaries in *NFPA 70E Tables 130.4(E)(a) and (b)* include an added safety margin to compensate for accidental movement of the worker. The interaction of exposed energized parts and test equipment or hand tools is the initiator of many shock and arc flash incidents when the tool or test lead becomes part of the circuit path. Some estimates state that 75% or more of arc incidents begin in this manner.

NFPA 70E Table 130.4(E)(a) identifies the shock protection approach boundaries to exposed energized parts of alternating current (AC) voltage systems. *NFPA 70E Table 130.4(E)(b)* identifies the shock protection approach boundaries to exposed energized parts of direct current (DC) voltage systems. For more information on limits of approach, see *NFPA 70E Informative Annex C*.

Know the Code

Electric Shock Protection Approach Boundaries to Exposed Energized Electrical Conductors or Circuit Parts for Alternating-Current Systems
NFPA 70E Table 130.4(E)(a)

Know the Code

Electric Shock Protection Approach Boundaries to Exposed Energized Electrical Conductors or Circuit Parts for Direct-Current Voltage Systems
NFPA 70E Table 130.4(E)(b)

Know the Code

Limits of Approach
NFPA 70E Informative Annex C

Figure 26 Worker using appropriate PPE.
Source: Honeywell | Salisbury

Always Use Insulated Tools

Insulated tools are intended to protect against providing an unintended path that becomes an arc fault. Anyone working close to energized parts must be extremely careful in their movements and use only insulated or insulating tools within the restricted approach boundary, along with the appropriate PPE.

2.2.0   Employer and Employee Responsibilities

It is everybody's responsibility to ensure that workplaces are free of electrical hazards. Employers must ensure their workers are properly trained and have the required PPE to perform the job safely. Employees must follow established procedures, stay within the job plan, inspect and use the appropriate PPE, and avoid shortcuts.

2.2.1 Employer Responsibilities

OSHA Standard 29, Part 1910 and *OSHA Standard 29, Part 1926* cover occupational safety and health standards. Within each standard are subparts that deal with specific areas of safety. For example, *OSHA Standard 29, Part 1910, Subpart S, Sections 331 through 335* address electrical safety-related work practices. *OSHA Standard 29, Section 1910.332* requires training for all employees exposed to electrical shock.

If individual OSHA standards do not address a specific safety issue, they are covered in the *Occupational Safety and Health Act of 1970 (OSH Act of 1970), Section 5(a)(1)*, which is often called the *General Duty Clause*. *OSH Act of 1970, Section 5(a) (1)* requires each employer to provide a workplace free from recognized hazards likely to cause death or serious physical harm. It also states that the employer must comply with all applicable occupational safety and health standards as determined by OSHA. NFPA 70E® was developed to provide specific methods that meet the safety requirements of OSHA. OSHA recognizes NFPA 70E® as a national consensus standard.

NOTE

Per NFPA 70E®, the first priority is to eliminate the hazard. PPE is considered the last line of defense against electrical hazards.

When OSHA compliance safety and health officers check a facility for compliance, they normally examine the safety training records and standard operating procedures (SOPs) or other company policies that ensure employees are properly trained. If there has been a reportable incident, especially an electrical incident, the compliance safety and health officers will want to see all training records, SOPs, work permits, and any other documents associated with the incident.

Employers are responsible for ensuring their workers are properly trained, have a safe work environment, are informed of hazards present in the work, have clearly written instructions, and are given all the necessary tools to do their jobs safely. Employers are also responsible for ensuring workers follow established safe work procedures and use all required PPE.

Part of an employer's responsibility involves having an electrical risk assessment completed on all electrical hazards in the workplace. Those analyses include a shock risk assessment and an arc flash risk assessment. After each hazard is analyzed and documented, PPE is identified for tasks performed on or near the hazard, and warnings are posted on the exterior of the equipment per *NFPA 70E Section 130.5(H)*. Labels should be readable from outside any boundary and must list the following data:

- Nominal system voltage
- AFB
- At least one of the following:
  - Available incident energy and the corresponding working distance, *or* the arc flash PPE category in *NFPA 70E Table 130.7(C)(15)(a)* or *NFPA 70E Table 130.7(C)(15)(b)* for the equipment
  - Minimum **arc rating** of clothing
  - Site-specific level of PPE

Figure 27 shows a typical warning label. The label lists the distances and exposure levels for both arc flash and shock and the recommended type of gloves to use. The equipment name and number identify the bus.

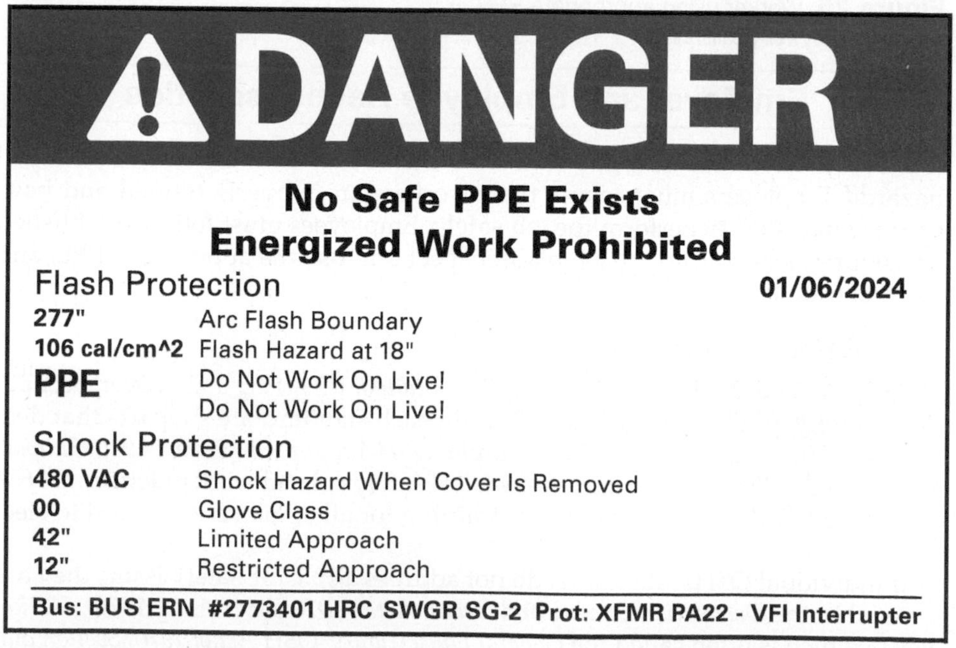

Know the Code

Equipment Labeling
NFPA 70E Section 130.5(H)

Know the Code

Arc Flash PPE Categories for Alternating Current (ac) Systems
NFPA 70E Table 130.7(C)(15)(a)

Know the Code

Arc Flash PPE Categories for dc Systems
NFPA 70E Table 130.7(C)(15)(b)

NOTE

A hazard assessment for other environmental hazards that may exist in the area should accompany the electrical risk assessment.

Arc rating: The maximum incident energy resistance demonstrated by a material (or a layered system of materials) prior to material breakdown or at the onset of a second-degree skin burn. Expressed in J/cm^2 or cal/cm^2.

⚠ DANGER

No Safe PPE Exists
Energized Work Prohibited

Flash Protection		01/06/2024
277"	Arc Flash Boundary	
106 cal/cm^2	Flash Hazard at 18"	
PPE	Do Not Work On Live!	
	Do Not Work On Live!	

Shock Protection	
480 VAC	Shock Hazard When Cover Is Removed
00	Glove Class
42"	Limited Approach
12"	Restricted Approach

Bus: BUS ERN #2773401 HRC SWGR SG-2 Prot: XFMR PA22 - VFI Interrupter

Figure 27 Electrical hazard warning label.

2.2.2 Employee Responsibilities

Employees are responsible for reading, understanding, and following all company safety policies. They must also inspect, use, and properly store all required PPE, look out for the safety of themselves and others, and work only within the job plan. Note that PPE must be inspected before each use.

In the past, plant electricians and maintenance workers were expected to simply address situations as they arose. That is no longer acceptable. *NFPA 70E Section 110.3(I)* requires job safety planning and a job briefing before starting any work involving exposure to electrical hazards. In addition, an Energized Electrical Work Permit (discussed later in Section 4.3.0) is required for any work involving exposure to electrical hazards per *NFPA 70E Section 130.2(A)*. Even a single electrical task, such as removing and replacing a motor starter, requires a specific job plan. Planning a job includes reviewing the risk assessment data associated with the equipment, reviewing the most recent power distribution drawings, ensuring the required PPE is available and in good condition, and obtaining a properly executed energized electrical work permit. In other words, you must have a specific goal in mind and perform only the planned job. Many electrical incidents can be traced to task creep. For example, if you take a current reading at a motor starter and notice a loose connection, do not grab your screwdriver and tighten it. You will make contact with an exposed energized part and greatly increase the likelihood of initiating an arc fault.

| 2.3.0 | **Human Performance and Workplace Electrical Safety** |

Human performance must be considered when conducting an electrical risk assessment. *NFPA 70E Informative Annex Q* provides information on human performance and workplace electrical safety. The principles of human performance are based on the following concepts:

- All people make mistakes.
- Error-prone situations and conditions are predictable, manageable, and preventable.
- Individual performance is influenced by organizational processes and values.
- High levels of worker performance are the result of positive reinforcement from leaders, peers, and subordinates.
- Incidents can be avoided by identifying why mistakes occur and applying the lessons learned from past incidents.

2.3.1 Human Performance Modes

Analyzing human performance helps to pinpoint potential error points. Workers operate in one or more human performance modes: rule-based, skill-based, and knowledge-based. These modes and the errors associated with them are described in *NFPA 70E Informative Annex Q.4* and are summarized as follows:

- *Rule-based mode* — Workers operate in rule-based mode when the task has been done before and follows written or practiced rules or procedures. Common errors in this mode include deviating from an approved procedure or applying the correct procedure to the wrong situation. Due to the predictable and tested rules and outcomes, the rule-based mode is the most desirable.

CAUTION

Always keep a clean work area when installing or removing equipment. Any scrap wire or even insulation could cause a fault or prevent protective devices from tripping. One cause of electrical faults is loose or uninstalled fasteners (bolts and screws) and tools left inside an electrical enclosure. Be sure to properly install doors and covers and account for all parts and tools. Do not leave a piece of electrical equipment unless all bolts and fasteners have been installed. Never use an electrical enclosure as a storage locker for spare parts and tools. Also, do not use the top surfaces of enclosures as storage or work areas, and make sure to maintain the required working space in front of the equipment per *NEC Section 110.26*.

Know the Code

Job Safety Planning and Job Briefing
NFPA 70E Section 110.3(I)

Know the Code

When Required
NFPA 70E Section 130.2(A)

Know the Code

Human Performance and Workplace Electrical Safety
NFPA 70E Informative Annex Q

Know the Code

Human Performance Modes and Associated Errors
NFPA 70E Informative Annex Q.4

- *Knowledge-based mode* — Workers operate in knowledge-based mode when a task has not been encountered before, and there is no specific procedure to follow. In this case, the worker must rely on knowledge and past experience to make decisions. The most common errors in this mode occur when decisions are based on an inaccurate assessment of the situation or focus on a single aspect of the problem rather than the big picture.
- *Skill-based mode* — Workers operate in skill-based mode when performing common and familiar tasks (e.g., operating a low-voltage circuit breaker). Common errors in skill-based mode are due to a lack of attention and a perceived reduction in risk.

2.3.2 Error Precursors

Error precursors are situations that put a worker at risk due to the demands of the task, conditions, worker attitude, and/or environment. *NFPA 70E Informative Annex Table Q.5* groups error precursors into the following categories:

- *Task demands* — High workload, monotony, time pressure, multiple tasks, critical/irreversible tasks
- *Work environment* — Distractions, interruptions, workarounds, unexpected conditions, new routine
- *Individual capabilities* — New task or technique, unsafe attitude, poor communication skills
- *Human nature* — Stress, habits, assumptions, complacency, overconfidence, shortcuts

Each of these precursors can be countered with one or more human performance tools.

2.3.3 Human Performance Tools

Human performance tools reduce the likelihood of error when applied to error precursors. *NFPA 70E Informative Annex Q.6* lists the following human performance tools:

- *Job planning and pre-job briefing* — These meetings identify each worker's role in the execution of the tasks.
- *Jobsite review* — A jobsite review is used to identify hazards and potential barriers or delays. It can be performed any time prior to or during work.
- *Post-job review* — A post-job review provides feedback that can be applied to future jobs.
- *Procedure use and adherence* — The worker must read and fully understand each step in a procedure before attempting to complete it. It is helpful to track progress by checking off each step as it is completed. This verifies that each step has been done in the specified sequence and ensures that the procedure can be resumed at the correct point if interrupted. If a procedure cannot be used as written or the expected result cannot be predicted, stop and resolve any issues before proceeding.

Know the Code

Error Precursor Identification and Human Performance Tool Selection (see Q.5 and Q.6.1)
NFPA 70E Informative Annex Table Q.5

Error precursors: Situations that increase risk to a worker due to demands of the task, environmental conditions, and/or human error.

Know the Code

Human Performance Tools
NFPA 70E Informative Annex Q.6

- *Self-check with verbalization* — A verbal self-check is a valuable tool when performing a critical or irreversible procedure. It is also known by the acronym STAR: Stop, Think, Act, and Review. Before, during, and after performing a task that cannot be reversed, the worker should stop, think, and verbalize the intended action.

- *Three-way communication* — When a statement is made by a sender, it is then repeated by the receiver to confirm the accuracy of the message, and again validated by the sender. Whenever possible, letters should be stated using the phonetic alphabet. For example, say "alpha" rather than "a," "bravo" rather than "b," and so on.

- *Stop when unsure* — When a worker is unable to follow a procedure as written, if an unexpected event occurs, or if the worker has a feeling that something is not right, then the worker must stop and obtain further direction before proceeding. All workers must be trained to recognize that phrases such as "I think" or "I'm pretty sure" are dangerous when it comes to safety.

- *Flagging and blocking* — Flagging is used to mark, label, or otherwise identify components to ensure the correct component is operated at the right time under the required conditions. Flags are used when operating look-alike equipment, working on multiple components, performing frequent operations in a short period of time, or interrupting process-critical equipment. Blocking is a method of physically preventing access to an area or equipment controls (e.g., barricades, fences, or hinged covers on switches or control buttons). Blocking can be used in conjunction with flagging.

The reduction or elimination of electrical incidents requires that all members of an organization work together to promote a culture that values error prevention and the use of human performance tools to identify and prevent error-prone situations and conditions.

2.0.0 Section Review

1. NFPA 70E® requires special safety procedures when working on or near circuits having AC voltage levels of *more* than _____.
   a. 24 V line-to-line
   b. 50 V line-to-line
   c. 120 V line-to-line
   d. 240 V line-to-line

2. The OSHA clause that requires each employer to provide a workplace free from recognized hazards is known as the _____.
   a. General Duty Clause
   b. Workplace Safety Clause
   c. Hazard Prevention Clause
   d. Employee Rights Clause

3. According to *NFPA 70E Informative Annex Q*, which of the following is *true* regarding human performance?
   a. Some people never make mistakes.
   b. Individual performance is rarely influenced by organizational values.
   c. High levels of worker performance are the result of positive reinforcement from peers, not leaders or subordinates.
   d. Error-prone situations and conditions are predictable, manageable, and preventable.

3.0.0 Electrical Incidents and Prevention

Performance Tasks

There are no Performance Tasks in this section.

Objective

Identify the causes of electrical incidents and explain how they can be prevented.

a. Identify the causes of electrical incidents.
b. Identify safety-related work practices.
c. Describe the personal protective equipment (PPE) used to protect against electrical hazards.
d. Describe the other tools and protective equipment used to protect against electrical hazards.

Calorie: The amount of heat energy required to raise the temperature of 1 gram of water by 1°C.

An electrical incident is a violent occurrence involving electrical power or energy. The overwhelming majority of electrical incidents are preventable.

NFPA 70E® uses the term *incident energy* to describe the amount of electrical energy applied to a surface positioned at a specific distance from the source during an electrical arc event. A measurement unit used to describe thermal energy transfer and incident energy is calories per centimeter squared (cal/cm^2). A **calorie** is the quantity of heat required to raise the temperature of 1 gram of pure water by 1°C (slightly less than 2°F). The term calorie has now been superseded by the joule in scientific usage. A calorie is about 4 joules. As a result, you will often see PPE ratings with two numbers, one for joules and one for calories.

Electrical injury is best prevented by simply avoiding electrical hazard exposure. If avoidance is not an option, then the proper selection and use of PPE is required, along with the use of the proper safe work practices applicable to the job. When an electrical incident or near miss occurs, the root cause must be determined to prevent a recurrence. The investigation should answer the following questions:

- What actions led up to the incident? At what point(s) could steps have been taken to prevent it?
- Why was work performed on energized equipment?
- What procedures or safe work practices were already in place that should have prevented this incident?
- Was the worker properly trained and supervised?
- Was the worker provided with all required PPE?
- If all preventive actions were taken and all safe work practices followed, did the PPE fail?
- Should the procedure be changed to include additional preventive actions and/or PPE?

3.1.0   Causes of Electrical Incidents

There are three basic causes of electrical incidents: unsafe conditions, unsafe equipment, and unsafe acts. Safety regulations are designed to prevent unsafe equipment and conditions. In addition, employer awareness and new technologies, ground fault circuit interrupters (GFCIs), touch-safe (encased) terminals, and insulated tools have greatly reduced incidents due to unsafe conditions and equipment.

So why do electrical incidents continue to happen? Most electrical injuries result from failure to follow safe work practices and not being aware of or ignoring a hazard. OSHA lists the following activities as some of the most frequent causes of electrical injuries:

- Failure to place the circuit or equipment in an ESWC
- Contact with power lines by ladders, powered construction equipment, earth-moving equipment, and construction tools such as long-handled cement finishing floats

- Lack of ground fault protection on outlet receptacles, powered hand tools, extension cords, and installed electrical equipment

- Damaged equipment such as cut, nicked, or pinched power cords and cables, worn insulation on power cords or cables, missing ground prongs, and damaged tool casings

- Path to ground missing or interrupted due to loose or broken equipment grounding conductors, improperly grounded equipment, improper grounds, or extremely dry conditions around existing grounding electrodes

- Equipment not being used in the prescribed manner, such as fabricating extension cords from multi-receptacle boxes or nonmetallic (NM) sheathed cable, using power tools with modified cords, and using oversized fuses or circuit breakers

3.2.0   Safety-Related Work Practices

The safety-related work practices described in *NFPA 70E Article 110* are implemented by the employees. The employer specifies the safety-related work practices and trains the employees. The following are a few safe work practices that apply to working on or near electrical equipment:

- Distractions are dangerous. Avoid headphones, earbuds, cell phones, and other sources of noise.

- Always assume that electrical equipment is energized until you personally verify that it is in an ESWC. Note that the verification process itself is considered energized work.

- Never wear jewelry or wire-rimmed glasses when working on or near energized electrical equipment.

- Plan every job in accordance with current company policies.

- Be sure you understand how to operate the equipment.

- Inspect PPE before each use.

- Inspect tools and portable equipment before use.

- Be completely sure of the boundaries associated with the job to be performed.

- Before entering any boundary, make sure that you and any coworkers are properly suited with the required PPE.

- Ensure there is adequate lighting for the task.

- Visually inspect the equipment on which the work must be performed, including any attached or nearby equipment.

- Do not reach blindly near energized equipment.

- Perform only the planned work.

- Exit the equipment and make sure no hardware or trash is left at the worksite.

- Only bring voltage-related tools and equipment into the boundary.

A pre-job briefing is required before each job. The team may be made up of only two people, but going over the plan ensures that everybody understands the work to be performed and the hazards involved. After the job is completed, a post-job briefing should be held so that the work can be reviewed and properly documented. If changes are needed in future job plans or procedures, they are submitted at this time.

Know the Code

General Requirements for Electrical Safety-Related Work Practices
NFPA 70E Article 110

Tool Control

Some companies use a tool log to inventory the tools in use on each job. This helps to ensure that no tools are inadvertently left in the equipment when the job is completed.

3.3.0 Personal Protective Equipment and Tools

PPE is designed to protect specific body parts. PPE is selected based on the electrical shock risk assessment and the arc flash risk assessment.

Some apparel and accessories can increase the risk of an incident occurring or the extent of personal injury. Do not wear loose-hanging clothing, exposed metal buckles, watches, or jewelry when working on or near electrical equipment. Wear clothing that covers the hands, arms, legs, and body. Be sure that shirttails are tucked and sleeves are unrolled, buttoned, and tucked into gloves. Tie back loose hair.

Per *NFPA 70E Section 130.7(C)(11)*, where employees are exposed to electrical hazards, clothing must be made of arc-rated material such as flame-retardant cotton. When exposed to extreme heat, flammable synthetic clothing melts to the skin, causing further injury. Although arc-rated material can ignite, it does not continue to burn after the source of ignition is removed.

Specialty clothing, such as high-visibility vests, are available as arc-rated but are not required to meet the exposure level of the base clothing.

> **WARNING!**
>
> Certain types of common construction clothing, such as standard high-visibility vests, cold weather gear, head scarves, etc., do not protect against arc flash and may not be worn within the AFB. In the event of an arc flash, these items will melt or burn and may reduce the protection of any arc-rated clothing worn underneath them.

Arc-rated clothing must have an **arc thermal performance value (ATPV)** or **energy breakopen threshold (EBT)** rating that provides adequate protection against the expected hazards. These values are determined by how long the material can withstand thermal energy before there is a 50% probability that it will cause a second-degree burn (ATPV) or develop a hole or other opening that would allow flames to pass through the material (EBT). Footwear must be non-conductive and high enough to protect the ankle area not covered by pants. Many companies provide arc-rated coveralls that can be worn over regular natural fiber clothing. In some locations, arc-rated clothing is standard daily wear. In other locations, arc-rated clothing and other PPE must be readily available when required.

3.3.1 Arc Flash Protective Clothing and Equipment

NFPA 70E Section 130.7(C) covers PPE. *NFPA 70E Tables 130.7(C)(15)(a) and 130.7(C)(15)(b)* list the PPE category classifications for various tasks. *NFPA 70E Table 130.7(C)(15)(c)* details the protective clothing and PPE required for Categories 1 through 4. *Figure 28* shows various types of PPE.

A wraparound arc-rated face shield with a chin protector attached to a hard hat may provide adequate protection for the neck, head, and face areas for PPE Category 1 and 2 exposures. Combining the face shield with an arc-rated balaclava (*Figure 29*) can provide protection for anticipated exposures under 12 cal/cm^2 (50.2 J/cm^2). Where the exposure is expected to exceed 12 cal/cm^2 (50.2 J/cm^2), an arc-rated hood must be used per *NFPA 70E Section 130.7(C)(10)(b)(2)*. Hearing protection must be the insert type. Although not offered as arc-rated components, conventional earplugs have not been shown to melt or cause increased injury.

Know the Code

Clothing Material Characteristics
NFPA 70E Section 130.7(C)(11)

Arc thermal performance value (ATPV): The incident energy limit that a flame-resistant material can withstand before it breaks down and loses its ability to protect the wearer. Expressed in J/cm^2 or cal/cm^2.

Energy breakopen threshold (EBT): The incident energy limit that a flame-resistant material can withstand before the formation of one or more holes that would allow flames to penetrate the material. Expressed in J/cm^2 or cal/cm^2.

Know the Code

Personal Protective Equipment (PPE)
NFPA 70E Section 130.7(C)

Know the Code

Arc Flash PPE Categories for Alternating Current (ac) Systems
NFPA 70E Table 130.7(C)(15)(a)

Know the Code

Arc Flash PPE Categories for dc Systems
NFPA 70E Table 130.7(C)(15)(b)

Know the Code

Personal Protective Equipment (PPE)
NFPA 70E Table 130.7(C)(15)(c)

Gloves Must
Be Rated

Leather Glove
Protectors

**Face Shield with Chin
Protector and Hard Hat**

Flash Suit

Gloves over
Sleeves

3 | CAL/CM²

Hood

Pants over
Boots

Shoe Covers

Kit Suit

Figure 28 Various types of PPE.
Source: Honeywell | Salisbury

NFPA 70E Section 130.7(C)(9) provides requirements for the selection of protective clothing. It discusses layering, use of flammable clothing, and elastics.

NFPA 70E Section 130.7(C)(10) explains arc flash protective equipment, including arc flash suits and the face, hand, and foot protection required when an arc hazard is present.

Know the Code

Factors in Selection of Protective Clothing
NFPA 70E Section 130.7(C)(9)

Know the Code

Arc Flash Protective Equipment
NFPA 70E Section 130.7(C)(10)

Figure 29 Balaclava (sock hood).
Source: Honeywell | Salisbury

NFPA 70E Section 130.7(C)(12) identifies clothing and other apparel that are not permitted when an arc flash hazard is present. These materials should not be worn by electrical workers at any time.

NFPA 70E Section 130.7(C)(13) lists the requirements for the inspection, care, and maintenance of arc-rated clothing and arc flash suits. Each rated garment has an ATPV or EBT rating on its tag. The tag also provides instructions for washing the garment. Many manufacturers recommend home laundering using a mild detergent. Home laundering is recommended since commercial laundering may degrade the protective characteristics of the material. The use of fabric softeners is generally not recommended.

Check with the manufacturer to determine the expected lifespan of the arc rating. Some protective clothing is designed for a single use, while others retain their rating for the lifetime of the garment with home laundering.

WARNING!

Garments with holes or rips are not reliable. Garments with sleeves rolled up or fasteners undone will not provide overall protection.

All arc-rated PPE apparel must be inspected prior to each use. Look for tears, worn spots, and contamination such as oil or grease. If PPE is contaminated or damaged, do not use it. Turn it in for cleaning, repair, or replacement.

Note that arc-rated clothing may be repaired using the same material as the clothing. When labels, embroidered emblems, or logos are attached to arc-rated clothing, Informational Note No. 2 in *NFPA 70E Section 130.7(C)(13)(d)* requires that these items be affixed in accordance with ASTM F1506, *Standard Performance Specification for Flame Resistant and Electric Arc Rated Protective Clothing Worn by Workers Exposed to Flames and Electric Arcs*. For other PPE standards, see *NFPA 70E Table 130.7(C)(14)*. The selection of PPE required for various tasks is covered under *NFPA 70E Section 130.5(G)*, which uses the incident energy analysis method, and *NFPA 70E Section 130.7(C)(15)*, which uses the arc flash category method. Equipment labels can only cite one of the two methods.

In the past, electrical arc flash protective equipment had a reputation for being hot, heavy, and uncomfortable, restricting both movement and vision. In addition, the selection of protective apparel far above the extent of the hazard was a factor in incidents caused by low visibility or lack of manual dexterity. An obvious danger of uncomfortable or overprotective PPE is that workers avoid using it. The comfort index of arc-rated apparel has improved greatly due to the use of lighter materials, increased breathability, and higher insulating values. For example, when comparing the face shield of a 10-year-old arc flash hood with one of the same class purchased last year, light transmission and visibility are greatly improved (*Figure 30*).

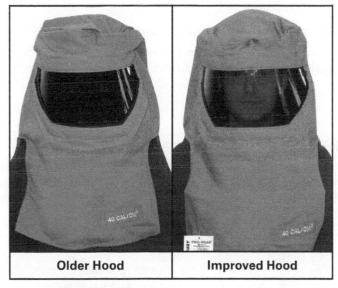

| Older Hood | Improved Hood |

Figure 30 Newer arc hoods offer improved comfort and visibility.
Source: Honeywell | Salisbury

Arc-rated garments may be layered for additional protection. See *NFPA 70E Informative Annex M*.

NFPA 70E Section 130.7(D) identifies requirements for other protective equipment including insulated and insulating tools, ropes, rubber insulating equipment, protective shields, and physical or mechanical barriers that are likely to be used in situations involving exposure to electrical hazards.

NFPA 70E Table 130.7(E) identifies standards for other protective equipment. Most of these standards cover manufacturing requirements such as ASTM D1048, *Standard Specification for Rubber Insulating Blankets*. These standards identify equipment ratings within the standard, the markings that identify equipment conforming to the standard, and manufacturer testing requirements. However, these standards do not provide specific requirements for user inspection and testing. Other standards address requirements for in-service care, inspection, and testing, such as ASTM F479, *Standard Specification for In-Service Care of Insulating Blankets*.

Know the Code

Standards for PPE
NFPA 70E Table 130.7(C)(14)

Know the Code

Incident Energy Analysis Method
NFPA 70E Section 130.5(G)

Know the Code

Arc Flash PPE Category Method
NFPA 70E Section 130.7(C)(15)

Job Preplanning

Before approaching a task wearing a flash suit and hood, put on all required PPE and practice working in similar lighting and space conditions in a safe area. Try using the tools you will need for the job while wearing the gloves and their protectors. Think about the location of the planned job. Is it a tight work area? Is there adequate lighting? If not, additional illumination must be provided per *NFPA 70E Section 130.8(C)*.

NOTE

Any layering combination must be tested by an approved laboratory. Suppliers of arc-rated materials or garments often have layering information available on request.

Know the Code

Layering of Protective Clothing and Total System Arc Rating
NFPA 70E Informative Annex M

Know the Code

Other Protective Equipment
NFPA 70E Section 130.7(D)

Know the Code

Standards for Other Protective Equipment
NFPA 70E Table 130.7(E)

3.3.2 Ratings of Rubber Insulating Gloves

Gloves must be rated for different voltages. *Table 2* shows the glove classes, DC voltage ratings, and color codes for gloves meeting the requirements of ASTM D120, *Standard Specification for Rubber Insulating Gloves*.

TABLE 2 ASTM DC Voltage Rating of Gloves

Class of Glove	Proof-Test Voltage	Minimum Breakdown Voltage	Color Specification
00	10,000	13,000	Beige
0	20,000	35,000	Red
1	40,000	60,000	White
2	50,000	70,000	Yellow
3	60,000	80,000	Green
4	70,000	90,000	Orange

Gloves in the original standard included Classes 1, 2, 3, and 4. Classes 00 and 0 are relatively new and used at 1,000 V and below. The higher the class number, the higher its voltage rating. The voltage rating is marked on a color-coded label near the wrist of the glove. The glove color itself does not identify the voltage class.

3.3.3 Selecting, Inspecting, Using, and Maintaining PPE

NFPA 70E Section 130.7(B) requires protective equipment to be maintained in a safe and reliable condition, stored properly, and visually inspected before each use.

NOTE

The use of glove Classes 00 through 4 has no relation to the PPE categories often used to classify levels of arc flash hazard.

Know the Code

Care of Equipment
NFPA 70E Section 130.7(B)

PPE Is Not Your Best Line of Defense

Many people hear "electrical safety" or "hazard" and automatically think of PPE. That is understandable because you have been trained to use PPE to protect yourself. However, PPE is not your first line of defense against electrical hazards. PPE comes into play only when all other possible measures have been taken to eliminate the need to be exposed to an electrical hazard. PPE like this is your last line of defense against injury or death.

Source: Oberon

The requirements of the national consensus standards are incorporated into the requirements of OSHA and NFPA 70E®. As stated previously, specific requirements for the inspection, testing, and maintenance of various types of PPE are provided in individual ASTM specifications. For example, ASTM F496, *Standard Specifications for In-Service Care of Insulating Gloves and Sleeves*, defines the maximum electrical test intervals, visual inspection requirements, and cleaning methods for this equipment. For PPE not covered by a specific standard, consult the manufacturer's literature supplied with the equipment.

Properly rated rubber insulating equipment is used to protect specific body part(s) against electrical shock hazards. The OSHA requirements for rubber insulating protective equipment are found in *OSHA Standard 29, Section 1910.137*. All rubber insulating equipment must be electrically tested before first issue and at the maximum intervals listed in *NFPA 70E Table 130.7(C)(7)(b)*. During testing of rubber equipment, each item is marked with a unique identification number, electrical test date, and expiration date. Pay close attention to both dates.

Visually inspect rubber insulating equipment before use each day and whenever damage is suspected. Remove rubber gloves from the protective leathers and inspect both individually. Look for any grit or wire bits that can damage the gloves. *Figure 31* identifies common problems to look for during inspection. Turn rubber gloves and leathers inside out during inspection. Lightly stretch rubber blankets and gloves to look for small cracks or checking. Rubber blankets are stretched by laying them flat, folding them diagonally, and rolling in each direction on both sides of the blanket.

Know the Code

Rubber Insulating Equipment, Maximum Test Intervals
NFPA 70E Table 130.7(C)(7)(b)

Cracking and Cutting

This type of damage is caused by prolonged folding or compressing.

UV Checking

Storing in areas exposed to prolonged sunlight causes UV checking.

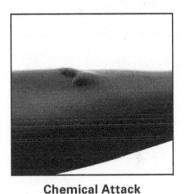

Chemical Attack

This photo shows swelling caused by oils and petroleum compounds.

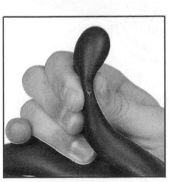

Snags

Damage shown is due to wood or metal splinters or other sharp objects.

Avoid Folding Electrical Gloves

The strain on rubber at a folded point is equal to stretching the glove to twice its length.

Avoid Storing Inside Out

Storing reversed gloves strains the rubber severely and promotes ozone cutting.

Figure 31 Glove inspection.
Source: Honeywell | Salisbury

Figure 32 Rubber glove on a glove inflator.
Source: Honeywell | Salisbury

Know the Code

Electric Shock Protection
NFPA 70E Section 130.7(C)(7)(a)

CAUTION

Do not wear watches, rings, or bracelets under rubber gloves. Sharp edges can cause punctures.

CAUTION

Do not use talcum powder in gloves. Some powders contain products that can damage the rubber.

Rubber gloves must also undergo an air test before each use and whenever damage is suspected. The gloves may be inflated with a portable inflator (*Figure 32*) or by flipping the glove and capturing air by rolling the cuff upward. Inflate gloves to no more than 150% of their normal size. Air leaks may be detected by feeling for leakage against your cheek or looking for dust puffs when glove dust is used. On a jobsite, it may be necessary to find a sheltered area in which to perform the air test.

WARNING!

When leakage is suspected, immediately remove the gloves from service. Do not use them.

Gloves and other rubber insulating materials can be damaged by petroleum products and hand lotions. Some types are also affected by sunlight and ozone. When rubber insulating equipment is contaminated by oils or grease, wipe off the contamination immediately. As soon as practical, wash the item with an approved cleaning product, such as mild dishwashing liquid.

Damage to rubber insulating products can be avoided by carefully following storage procedures. Rubber insulating equipment must not be folded or stored inside out. Blankets (*Figure 33*) should be rolled and stored in a blanket rollup or canister intended for that purpose. Store clamp pins in a separate container. Gloves issued for use should be stored inside protective glove bags inside their leathers with cuffs down.

WARNING!

Ensure that rubber insulating gloves and leathers fit the hand properly and that the leathers are matched to the classification of the glove. Failure to do so will not provide the expected level of protection.

Rubber Blanket with Storage Rollup

Figure 33 Rubber blanket.
Source: Honeywell | Salisbury

Gloves are intended for use only under their protective leathers. When a task requires dexterity not possible with leathers on, the leathers may be removed only to complete that portion of the task. Per *NFPA 70E Section 130.7(C)(7)(a)*, rubber insulating gloves shall be permitted to be used without leather protectors as long as there is no activity performed that risks cutting or damaging the glove, the gloves are electrically retested before reuse, and the voltage rating of the gloves is reduced by 50% for Class 00 and by one whole class for Classes 0 through 4.

Gloves do not breathe, causing dampness inside. Glove dust or cotton liners can be used to increase wearer comfort.

3.4.0 Other Tools and Protective Equipment

If your job requires the use of insulated or insulating hand tools, live-line tools (insulated hot sticks with various tools attached), temporary grounding devices, grounding mats, or other protective equipment, you must be trained in the use, inspection, cleaning, and storage of these items. *NFPA 70E Section 130.7(D)* covers requirements for the use of insulated tools and other protective equipment. National consensus standards for the care and use of other protective equipment are identified in *NFPA 70E Table 130.7(E)*.

3.4.1 Insulated and Insulating Tools

Insulated and insulating tools are required whenever work is performed within the restricted approach boundary of circuits rated 1,000 V and below. These tools are made of or covered with dielectric material that protects against initiation of an electrical arc by accidental contact with an energized part or ground.

Insulated tools have an outer coating of insulating material while insulating tools are constructed primarily of insulating material with metal inserts only at the working ends or for reinforcement. These tools must conform to ASTM F1505, *Standard Specification for Insulated and Insulating Hand Tools*. This standard requires that insulated and insulating tools be:

- Rated for up to 1,000 V (AC or DC)
- Flame-resistant
- Coated with one or more layers of insulation using contrasting colors for multiple layers
- Intended for use at temperatures between –20°C to 70°C or marked with the letter C if intended for use at –40°C
- Marked with the manufacturer's name or trademark, a double triangle, and the year of manufacture
- Visually inspected before use each day

Tools that show separation, cracking, or embedded contamination of insulation cannot be used for work within the restricted approach boundary.

WARNING!

Use only labeled tools complying with ASTM F1505 or other accepted standards. Taping, shrink tubing, or plastic dipping of tools will not provide the same level of protection as properly rated tools.

3.4.2 Live-Line Tools

Requirements for live-line tools are found in *OSHA Standard 29, Section 1910.269(j)*. Tools made of fiberglass-reinforced plastic (FRP) must be designed and constructed to withstand a test voltage of 100,000 V per foot of length for five minutes according to ASTM F711, *Standard Specification for Fiberglass-Reinforced Plastic (FRP) Rod and Tube Used in Live-Line Tools*. Live-line tools must be:

- Wiped clean and visually inspected before use for any defects that could affect the insulating quality or mechanical integrity of the tool
- Removed from service for electrical testing and additional inspection if any defect or contamination is visible after wiping
- Removed from service at least once every two years for electrical testing and additional visual inspection
- Handled with care
- Stored in a protective sleeve or tube to prevent damage and protect against condensation and dust

Know the Code

Other Protective Equipment
NFPA 70E Section 130.7(D)

Know the Code

Standards on Other Protective Equipment
NFPA 70E Table 130.7(E)

Arc Suppression Blanket

This arc suppression blanket is made out of ballistic material that provides a barricade to deflect potential arc hazards. It can be attached using clips through the side loops.

Source: Honeywell | Salisbury

CAUTION

Tools stored in a cool, dry environment (such as an air-conditioned service van) must be allowed to acclimate to the work environment before use. Condensation on a tool may reduce or eliminate its insulating qualities.

3.4.3 Temporary Grounding Equipment

Temporary grounding jumpers (*Figure 34*) are required to carry the maximum fault current expected at a location for the time necessary for protective devices to operate to clear the fault. The clamps must be able to carry the expected fault current and create a mechanical connection strong enough to withstand the magnetic forces generated during a fault. Failure of the temporary grounding equipment or resistance in connections or conductors will expose workers to a lethal shock hazard in the event of a lightning strike or inadvertent reenergization of the circuit. The technical requirements for temporary grounding equipment are found in ASTM F855, *Standard Specifications for Temporary Protective Grounds to Be Used on De-energized Electric Power Lines and Equipment.*

WARNING!

In overhead distribution, conductors from an energized line may blow into or fall onto conductors of a de-energized line that may be on the same structure or crossing under it. Switching errors may also reenergize a circuit that has been tagged out.

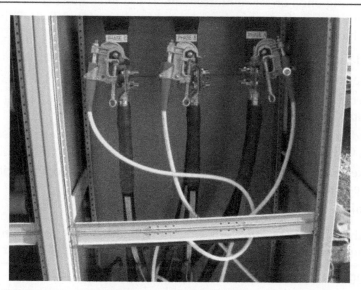

Figure 34 Temporary grounding jumpers in place on incoming feeders.
Source: Jim Mitchem

NFPA 70E Section 120.5(B)(7) addresses requirements for temporary grounding equipment in creating an ESWC. *NFPA 70E Section 250.4* covers visual inspection, testing, and storage of temporary grounding equipment. NFPA 70B, *Standard for Electrical Equipment Maintenance*, provides additional information on temporary grounding.

The application and removal of temporary grounds is considered work on energized parts and is hazardous. There are many instances of electrical shock and arc fault events during the installation or removal of temporary grounding. Procedures and required PPE must be used per the written policies of the employer.

Temporary grounds must be sized to limit the voltage potential across a worker's body to 100 V for a 15-cycle event or 75 V for a 30-cycle event. This is typical for work on overhead lines. It is critical that there be no added impedance in the ground path due to loose connections or surface contamination.

Temporary grounding is connected by first clamping one end of the jumper to ground and then clamping the other end of the jumper to line. Subsequent jumpers are connected from the ground point to each other, to line, or to phase. When temporary grounding is applied, each ground should be identified and logged by location. Removal is in reverse order, with the last connection removed being the initial connection to ground. Be sure to account for all ground jumpers before energization.

Know the Code

Grounding
NFPA 70E Section 120.5(B)(7)

Know the Code

Grounding Equipment
NFPA 70E Section 250.4

As with all protective equipment, temporary grounding equipment must be inspected before each use. When equipment is in place for more than one day, inspect the grounding equipment before each shift. Do not remove and replace it, as doing so creates an additional hazard.

Visual inspection includes examining for dirt or corrosion on contact surfaces and damage to conductor jackets. Also, check whether any conductor strands have pulled out at a ferrule and look for loose connections between the ferrule and clamp.

Store temporary grounds neatly rolled and hung up or placed in storage bags. Identify each temporary ground set or jumper to allow traceability of use and testing. Inspect grounding equipment periodically while in storage. Perform electrical testing at a minimum of once per year and after every instance of carrying fault current.

The testing may be done with a high current source or a micro-ohmmeter. The acceptance criterion is impedance through the clamps and conductor that will not result in excessive voltage potential across the worker during a fault.

Figure 35 shows live-line work using a hot stick to perform tests on an overhead line. Hot sticks can also be used to place temporary grounds on overhead lines. Doing so creates an ESWC for a line worker standing above the clamp-on grounding bar cluster. The line worker is not exposed to shock hazard because the worker's body is within an equipotential plane created through the bonding. The worker in *Figure 36* has created an equipotential plane between mobile equipment and the work location. The truck must be bonded to the pole ground, the conductive ground mat, and a temporary ground rod installed near the pole base. The grounding mat places the worker at the same electrical potential as the mobile equipment.

Figure 35 Hot stick in use.
Source: Rick Goodfriend

Figure 36 Grounding mat in use.
Source: A.B. Chance Co./Hubbell Power Systems

3.0.0 Section Review

1. Most electrical injuries result from failure to _____.
   a. wear appropriate PPE
   b. document procedures
   c. coordinate work activities
   d. follow safe work practices

2. Which of the following is *true* regarding a pre-job briefing?
   a. Pre-job briefings are not necessary on crews with only two people.
   b. A pre-job briefing is required on every job.
   c. Pre-job briefings are only required for complex tasks or when tasks are performed for the first time.
   d. Pre-job briefings must be incorporated into a safety document that is reviewed by OSHA.

3. The thermal energy level that a material can withstand before the formation of one or more holes is known as the _____.
   a. EBT
   b. ATPV
   c. AIC
   d. ESWC

4. According to ASTM F1505, *Standard Specification for Insulated and Insulating Hand Tools*, insulated and insulating tools *must* be rated up to _____.
   a. 240 V
   b. 480 V
   c. 600 V
   d. 1,000 V

4.0.0 Analyzing Electrical Hazards

Performance Task

1. Given a specific electrical task and circumstances, complete an Energized Electrical Work Permit request.

Objective

Explain the procedures for analyzing electrical hazards.
a. Identify the steps in a shock risk assessment.
b. Identify the steps in an arc flash risk assessment.
c. Complete an Energized Electrical Work Permit.

This section addresses how to analyze shock hazards and arc flash hazards so that employees are aware of the anticipated hazard, approach boundaries, and other protective measures required. The time to perform an electrical risk assessment is before an electrical work permit request is needed. Waiting to perform a risk assessment until just before a perceived need tends to cast the process as an obstacle to getting the work done instead of an essential step in managing the hazards.

An electrical risk assessment should provide recommended methods for minimizing employee exposure to electrical hazards. The following are a few possible recommendations:

- Restricting access to electrical rooms and other locations containing electrical equipment
- Applying minimum clothing standards for workers who may be exposed to electrical hazards
- Keeping doors and covers closed and secure on enclosures housing electrical equipment

- Ensuring that the boundaries of an ESWC are clearly identified and similar items are flagged so there is positive indication of whether a component is in an ESWC or not
- Requiring a very high standard to justify employee exposure to the hazards of suspect equipment

---

WARNING!

Equipment that is suspect due to age, lack of maintenance, or environment represents a great increase in both the risk of an incident happening and in the amount of energy released. This is due to overcurrent devices not opening or having a delayed clearing time. That increased risk and hazard cannot be accurately calculated or quantified and must be avoided whenever possible.

- Requiring permits to ensure that:
  - Work on energized circuits is never performed for convenience, but only when adequate justification exists that indicates the work must be performed under energized conditions.
  - Documented task procedures are used to minimize both the risk of initiating an arc and the exposure to the arc flash.
  - All participants know the scope of the work and the boundaries of the task.
  - Unqualified personnel are kept outside of both the established flash protection and shock protection boundaries.

NFPA 70E Informative Annex E identifies typical principles that should be part of an electrical safety program. These include:
- Inspecting/evaluating the electrical equipment
- Maintaining the electrical equipment's insulation and integrity
- Planning every job and documenting first-time procedures
- De-energizing, if possible
- Anticipating unexpected events
- Identifying and minimizing the hazard
- Protecting yourself and coworkers from electric shock, burn, blast, and other hazards due to the working environment
- Using the right tools for the job
- Assessing people's abilities
- Auditing these principles

Know the Code

Electrical Safety Program
NFPA 70E Informative Annex E

Applying these principles will help to prevent incidents and ensure a safer working environment for everyone.

When determining the type and location of electrical hazards in a workplace, first examine the power distribution drawings for the facility being evaluated (*Figure 37*). Many facilities have single-line drawings showing the distribution of electrical power throughout the facility. General arrangement drawings and as-built drawings show the facility's layout and specific locations such as service disconnects, switchgear rooms, MCCs, terminal cabinets, cable trays, and bus ducts. Potentially dangerous locations throughout the facility must be identified. Also important are operation and maintenance manuals for specific equipment and other engineering documents, including short-circuit studies and coordination studies.

The standard operating procedures of the facility should provide specific step-by-step instructions for LOTO of individual equipment and of systems or production lines within the facility. In addition, every facility should have a hazard-awareness training program for casual visitors or other nonemployees who will be in process areas. These may include suppliers, technical representatives, and outside contractors.

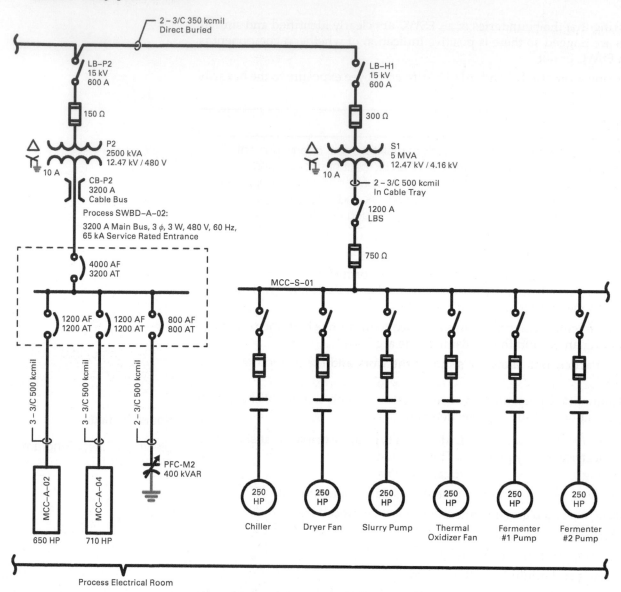

Figure 37 Simplified version of a single-line power distribution drawing.

4.1.0 Shock Risk Assessment

At its simplest level, a shock risk assessment identifies the presence of exposed energized parts, safe limits of approach for both qualified and unqualified personnel, and requirements for PPE. *NFPA 70E Section 130.4* discusses shock risk assessment and identifies approach boundaries to energized electrical conductors or circuit parts. A potential shock hazard exists when exposed parts, energized at 50 V or more, may be approached nearer than a safe distance by a person or object in contact with the person. *NFPA 70E Informative Annex C* provides additional information about approach boundaries to electrical hazards as presented in *NFPA 70E Tables 130.4(E)(a) and 130.4(E)(b)*. This discussion identifies the basis of the boundaries and the minimum air insulation distances required to prevent flashover.

A shock risk assessment should also identify the presence of concealed, covered, or guarded energized parts that would be exposed when a boundary or barrier is bypassed. It also addresses parts that are currently de-energized but could become energized by either intentional or unintentional means. Beyond identifying the potential electrical shock hazards, a shock risk assessment should also address methods to prevent both deliberate and inadvertent contact with energized parts.

When evaluating the possible shock hazards of electrical equipment, do not guess at anything. Use up-to-date drawings for the facility and equipment in question. Collect as much information as possible and ask questions. If documents or other information cannot be found, do not proceed until adequate information is available.

4.1.1 Knowledge

The first line of defense against electrical shock hazard is knowledge and the ability to recognize when you may be at risk. This knowledge allows you to comply with the technical aspects of risk assessment (i.e., what is the hazard, and what can I do about it?). Never work outside of your knowledge and comfort level. If you are uncertain or exposed to new systems, get assistance and don't proceed without a solid understanding. You may save your own life or that of a coworker.

Qualified personnel must be knowledgeable as to the construction and operation of the equipment in the work area. That equipment may be in a permanent facility, on a construction site, or in a temporary power situation. Qualified workers must be able to:

- Identify the electrical parts of the targeted machine or process
- Determine from drawings and associated documents which components are normally energized and what voltages are expected at each part
- Determine the operational status of equipment or conductors
- Identify potential sources of stored or backfeed energy
- Identify energy control points for LOTO and the methods of control
- Follow LOTO procedures for the company/facility and have access to the equipment required to implement those procedures
- Recognize energized parts that may be concealed

4.1.2 Variables

Another factor in risk assessment is looking beyond the technical data and results and examining factors both in and out of your control that can affect your ability to safely perform the task. First, examine the immediate environment. Is there adequate lighting? What about slip or trip hazards? Is it too noisy to permit communication while using hearing protection? Are you rushed, distracted, tired, fatigued, or not feeling well?

Also, ask "What if?" questions. What if you use a non-insulated tool and it slips? Is an observer needed to provide a warning if persons or equipment are near an approach boundary? What if you must adjust your working position? Could your arm or another body part cross the restricted approach boundary? Insulated cover-ups or rubber-insulated blankets and other rated products may be placed to insulate or guard energized parts. The risk assessment must identify all additional protective measures to be taken in addition to requirements for rubber insulating equipment.

4.1.3 Performing a Shock Risk Assessment

Perform the following steps to analyze the potential shock hazards associated with a given task.

Step 1 Determine the targeted electrical component or work area.

Step 2 Use single-line drawings, equipment nameplate data, and other available documents to determine the nominal voltages expected in the targeted work area.

Consider the Consequences

When work involving exposure to an electrical hazard is contemplated, first ask whether the circuit can be de-energized. If the answer is no, ask why not. If there seems to be adequate justification, ask if the work must be done now or whether it can be postponed until an outage may be taken. If a planned outage is not possible, ask if an unplanned outage and extended downtime due to equipment damage would be worse. Electrical incidents are relatively rare, but the cost when an incident does occur is unacceptable in terms of personal injury as well as property and consequential damages. Examine the damage caused by an arc blast and reconsider.

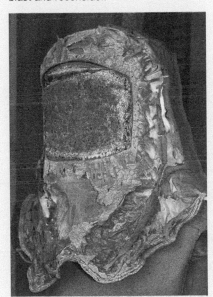

Source: Oberon

Paperwork

Many people believe that paperwork does not make them safer and just slows everything down. Paperwork does slow everyone down in a personal safety environment, but that is the point. By stopping what you are doing to fill out the permit and checklist, you are forced to consider how to complete the task safely. You must research the hazards, come up with a plan, and communicate the hazards and plan to everyone involved before you can safely proceed with the task. During this time, you may come to realize that the task you were about to perform is not safe and an outage is necessary.

Step 3 Refer to *NFPA 70E Tables 130.4(E)(a) and 130.4(E)(b)* to determine the shock protection boundaries for each voltage. Identify the following:

- Limited approach boundary
- Restricted approach boundary
- Means used to identify the approach boundaries (signs, barricades, or attendant)

Step 4 Review the proposed task and identify the locations of normally energized parts. This includes exposed energized parts and insulated, concealed, or guarded energized parts. Identify the scope of the planned work and consider whether it can be completed without crossing the restricted approach boundary. Select necessary PPE for shock protection. Determine if insulating materials will be used to guard exposed parts.

> **WARNING!**
>
> In some cases, such as systems rated 150 V and below, avoiding contact with exposed energized parts is the method used to protect against electrical shock. If this shock protection method is used, the task cannot allow any contact with exposed energized parts, including voltage testing.

Step 5 Determine any variables that could affect the work being performed in the targeted work area. Consider the possibility of inadvertent movement by the worker or an object in hand when selecting distance as the shock protection method.

Step 6 Document the results of the shock risk assessment. This data must be entered on the energized work permit form.

4.2.0 Arc Flash Risk Assessment

Know the Code

Arc Flash Risk Assessment
NFPA 70E Section 130.5

Know the Code

Incident Energy and Arc Flash Boundary Calculation Methods
NFPA 70E Informative Annex D

An electrical arc occurs any time the flow of electricity is initiated or interrupted. It can be caused by the opening and closing of a contact, connection to test equipment, or insertion or removal of circuit devices. Any time a minor arc is created, the potential exists for that arc to become an uncontrolled electrical arc flash due to environment or insulation breakdown. Electrical arc blasts have occurred when reclosing a circuit breaker after a trip or closing a switch into a fault. An uncontrolled electrical arc flash may also occur when the environment surrounding the equipment is disturbed. This disturbance may be from physical damage, moisture, dust, or even disturbed airflow. *NFPA 70E Section 130.5* explains when an arc flash risk assessment is required and describes how to determine an AFB and select required PPE. *NFPA 70E Informative Annex D* provides formulas for calculation of flash protection boundaries and estimates of incident energy exposure.

4.2.1 Purpose of Arc Flash Risk Assessment

Assessing arc flash hazards is more complicated than analyzing shock hazards. With a shock hazard, you are dealing with known voltage levels in a known environment, and a chart identifies the approach boundaries. With an arc flash, numerous variables control the intensity of the arc flash, including available fault current, required time to detect and clear an arcing fault, arc length, voltage, and distance from the fault.

An arc flash is an intense event. The extreme heat generated causes the air to expand rapidly, and the vaporization of copper results in additional hot gases causing more expansion. The instantaneous expansion of air creates a pressure wave similar to an explosion of dynamite. If the arc happens in an open space, such as an unenclosed switch or circuit breaker in a substation, the arc energy can dissipate in all directions. An arc of the same energy level occurring in an electrical enclosure or equipment results in the arc energy being concentrated, usually directly toward the worker at the equipment face.

The primary objective of an arc flash risk assessment is to identify when an arc flash hazard is expected to exist and the extent of the hazard. The first step is to identify the incident energy exposure and take any engineering approaches necessary to lower the incident energy determined by the assessment. After all engineering changes to limit the exposure have been implemented, it can then be evaluated for PPE.

The assessment establishes the AFB for a given electrical component or work area and identifies the PPE appropriate to protect the worker within that boundary against the onset of a second-degree burn. Per *NFPA 70E Section 130.7(C) (6)*, the **threshold incident energy level** for a second-degree burn is 1.2 cal/cm^2 (5 J/cm^2). Burns are classified according to the extent of cell damage as follows:

- First-degree burns are superficial burns, with only the outer layer of skin receiving damage. Indications are redness, mild swelling, tenderness, or pain.

- Second-degree burns are burns that penetrate the outer layer of skin and reach the inner layer of skin. Indications are the formation of blisters, swelling, seepage of fluids from the burn area, and severe pain.

- Third-degree burns are burns that penetrate all skin layers and reach the muscles and fatty tissues, causing cell death. The skin will appear charred, gray, waxy, or like dried-out leather. Victims may also have respiratory damage from inhalation of hot gas or copper vapor and may go into shock. They must receive immediate medical attention.

The use of apparel that is not arc-rated or inadequately rated increases the risk of burn injury. Synthetic or natural fiber garments burn and continue burning after the arc flash event is over.

Figure 38 shows the Rule of 9s for body percentage areas to determine how much of a person's body has been burned.

Know the Code

Body Protection
NFPA 70E Section 130.7(C)(6)

Threshold incident energy level: The amount of energy required to produce a second-degree burn (1.2 cal/cm^2 or 5 J/cm^2).

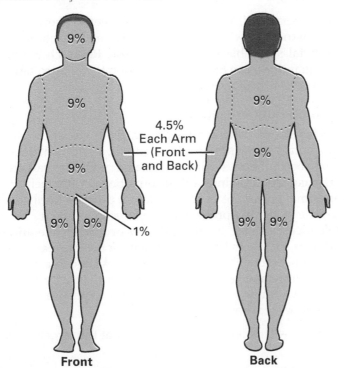

Figure 38 Body percentage areas (Rule of 9s).

Incident energy is the amount of heat received by a surface, such as unprotected skin, positioned at a given distance from an electrical arc flash. That incident energy is typically measured in calories per centimeter squared (cal/cm^2). An arc flash risk assessment determines the AFB and estimates/anticipates the level of PPE required for burn protection within that boundary. PPE is rated with an ATPV or EBT rating in cal/cm^2 or J/cm^2 that the garment can block or withstand before transferring enough energy to cause a second-degree burn (ATPV) or before developing a hole or other opening that would allow flames to pass through the material (EBT). *Figure 39* shows the calorie ratings of various types of PPE. It is important to ensure that all selected PPE meets or exceeds the minimum required calorie rating.

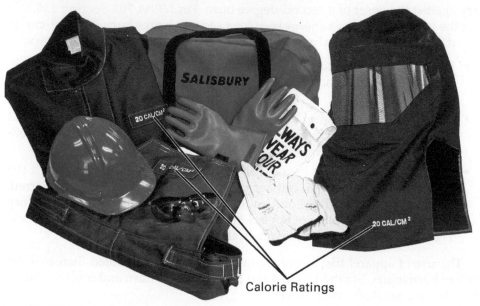

Calorie Ratings

Figure 39 PPE calorie markings.
Source: Honeywell | Salisbury

As previously discussed, in addition to the intense thermal energy present, the hazards associated with an arc flash event include shrapnel, vaporized gases, molten metal spray, blast effect, and noise. Research is underway that will attempt to quantify those hazards in addition to the burn hazard, but specific guidance is currently unavailable.

4.2.2 Information Requirements

As with a shock risk assessment, an arc flash risk assessment starts with the proper documentation (drawings, vendor manuals, equipment data, and system short-circuit or coordination studies associated with the targeted electrical equipment). Use current drawings and information for the facility and area in question. If information cannot be found, do not proceed until adequate information is available to identify protective measures to be taken.

This module focuses on electrical arc flashes occurring in equipment, sometimes referred to as *arc in a box*, as that is the most common arc situation faced by the industrial or commercial electrical worker. An arc in a box will generate higher incident energy and require a greater AFB than an arc in open air. To determine the AFB or anticipated incident energy exposure for a given situation, you must know the following:

- *Nominal voltage* — This represents the voltage at the targeted component or work area.
- *Short-circuit capacity (kA) of the electrical system supplying the targeted component* — In engineering analyses, this involves the utility supply source and all points of service, transformers, generators, and motors within the facility. The simpler methods discussed later in this module involve treating the work location as part of a simple radial distribution system served by the transformer immediately upstream.

- *Estimated time of arc flash exposure (in seconds)* — This depends on the settings and tolerance of the fault protective device(s) relied on to detect and respond to an arcing fault on the equipment (per cycle). When evaluating clearing times, consider both the trip time of protective relays and the operating time of circuit breakers or switches.

- *Working distance* — This is the distance from the worker's face and chest area to a prospective arc source. See *NFPA 70E Section 130.5(G)*. Personnel working at the face of a low-voltage MCC bucket will usually work about 18" from an arc flash inside that bucket.

4.2.3 Analyzing Electrical Arc Flash Hazards

NFPA 70E Section 130.5 provides requirements and methods for performing arc flash risk assessment. These methods cannot be used without knowledge of the electrical system capacity and characteristics of circuit protective devices over a probable range of fault current.

> **WARNING!**

The following discussion is relevant only to equipment believed to be properly maintained and in good operating condition so that protective devices are expected to work as intended per *NFPA 70E Section 210.6*. Poorly maintained or improperly set protective relays and circuit breakers may respond slowly, or not at all, to faults that should be cleared in the shortest possible time. Suspect equipment presents a level of uncertain performance that should result in an assessment. This assessment determines whether the degree of risk is unacceptable due to this increased incident energy that may be released.

NFPA 70E Informative Annex D contains detailed calculations to determine both the AFB and incident energy exposure at working distance. Other methods can be used, such as commercial software or *IEEE 1584, Guide for Performing Arc Flash Hazard Calculations*.

4.2.4 PPE Category Tables

Before using *NFPA 70E Tables 130.7(C)(15)(a) and 130.7(C)(15)(b)*, it is important to read and understand *NFPA 70E Article 130* in its entirety. *NFPA 70E Sections 130.5(F)(1) and (2)* identify two methods for selecting PPE and other equipment when it is determined that work will be performed within an AFB. The preferred method is to determine the anticipated incident energy exposure for each panel and location where electrical work may be done at that jobsite. This identifies the available arc energy that the PPE must be able to withstand. A label showing this required information must be applied.

NFPA 70E® provides an arc flash PPE category method using *NFPA 70E Tables 130.7(C)(15)(a), 130.7(C)(15)(b), and 130.7(C)(15)(c)*. This method is based on conservative assumptions of available fault current, maximum fault clearing time, and specified working distance. These tables do not quantify a task-specific anticipated incident energy exposure because they are derived from generic incident energy calculations. For example, for a 600 V Class MCC, the third row of *NFPA 70E Table 130.7(C)(15)(a)* indicates that Category 2 PPE must be used when there is no more than 65 kA of available fault current with an anticipated arc flash clearing time of no more than 0.03 seconds. If either value is greater, a detailed calculation of the anticipated incident energy exposure is required.

Once the PPE category has been determined, *NFPA 70E Table 130.7(C)(15) (c)* specifies the PPE required for the task. *NFPA 70E Informative Annex H* describes a simple two-category system that basically requires PPE Category 2 (minimum arc rating of 8) for daily work wear, adding additional layering to protect at PPE Category 4 (minimum arc rating of 40) when necessary.

Arc Energy and Working Distance

Small changes in work methods or position can have significant impacts in your exposure to injury from an arc flash. Since arc energy expands in all directions, its intensity changes as a square of the relative change in distance. For example, if your tools or vision require you to be at 14" instead of the typical 18" for work in a low-voltage MCC bucket, the potential arc energy hitting you will be 65% greater: $(18" \div 14")^2 = 1.65$. If you are using a 6" screwdriver to work at 24" ($18" + 6" = 24"$), the potential arc energy hitting you would be 44% smaller: $(18" \div 24")^2 = 0.56$.

Know the Code

Incident Energy and Arc Flash Boundary Calculation Methods
NFPA 70E Informative Annex D

Know the Code

Arc Flash PPE Category Method
NFPA 70E Section 130.7(C)(15)

Know the Code

Work Involving Electrical Hazards
NFPA 70E Article 130

Know the Code

Arc Flash PPE
NFPA 70E Sections 130.5(F)

Know the Code

Guidance on Selection of Protective Clothing and Other Personal Protective Equipment (PPE)
NFPA 70E Informative Annex H

As discussed previously, the best way to prevent electrical injury is to eliminate the hazard by de-energization and LOTO of electrical power sources. However, there are some circumstances in which worker exposure to an electrical hazard is unavoidable due to the design of the equipment or when de-energization would present a greater hazard. *NFPA 70E Section 110.2* lists conditions where work that involves exposure to electrical hazards may be justified. These include the following:

- *Additional hazards or increased risk* — The employer can demonstrate that de-energizing the circuit or equipment would introduce additional hazards or increased risk. Examples include life-support equipment, emergency alarm systems, and hazardous location ventilation equipment.

- *Infeasibility* — The employer can demonstrate that the task to be performed is infeasible in a de-energized state due to equipment design or operational limitations. Examples include equipment diagnosis and testing.

- *Equipment operating at 50 V or less* — The capacity of the source and any overcurrent protection between the energy source and the worker are considered, and it is determined that there will be no increased exposure to electrical burns or arcs.

Note that convenience, time, or costs are never reasons to work on energized equipment. Per *NFPA 70E Section 110.2*, normal operation is only permitted when the equipment is properly installed and maintained, the equipment is rated for the available fault current, all doors and covers are closed and secured, there is no evidence of impending failure, and it is being operated in accordance with its listing, labeling, and manufacturer instructions. Never operate equipment that fails to meet these conditions.

WARNING!

Per Exception 5 in *NFPA 70E Section 110.2(B)(7)*, circuits or circuit parts operating at less than 50 V to ground are not required to be de-energized where capacity or circuit protection limits energy to a level below that necessary to cause electrical burns or explosion from electric arcs. However, you need to be aware that work on these energized circuits can lead to a fault or trip of a low-voltage power, control, or signal circuit. This may disrupt the line in a process facility or result in failure of emergency equipment and systems. Battery systems of less than 50 V may have enough current to cause thermal burns in the event of a fault between positive and negative polarities. Be sure to assess all possible hazards before performing work of any kind.

NFPA 70E Section 130.2 requires that work involving exposure to electrical hazards be justified, planned, and authorized through a written Energized Electrical Work Permit. An example of an Energized Electrical Work Permit can be found in *NFPA 70E Informative Annex J.* Figure J.2 in *NFPA 70E Informative Annex J* provides a flow chart to determine whether an Energized Electrical Work Permit is required. Similar charts are used in many manufacturing and process facilities with dedicated electrical maintenance personnel.

Know the Code

Electrically Safe Work Condition
NFPA 70E Section 110.2

Change is Good

Although many electrical workers in the past considered "working hot" to be what good electricians can do, the culture is changing to recognize that good electricians avoid needless exposure to electrical hazards. In addition, many traditional practices in testing and troubleshooting present much greater hazard potential than alternate methods that are equally convenient. For example, measuring voltage in a transformer enclosure is unnecessary when the equipment served contains functional metering equipment.

Know the Code

Energized Electrical Work Permit
NFPA 70E Section 130.2

Know the Code

Energized Electrical Work Permit
NFPA 70E Informative Annex J

NOTE

NFPA 70E Section 130.2(C) identifies circumstances where activities involving exposure to electrical hazards may be allowed without an Energized Electrical Work Permit. However, many employers require that the permit process be used to justify, plan, analyze, authorize, and document all work involving exposure to electrical hazards. This provides an additional control to limit optional exposure.

4.3.1 Completing Part I

The permit process begins in Part I with a description of the circuits and equipment involved, the scope of work to be performed, and the justification of why the work must be performed under energized conditions. The first part of the permit will probably be completed by the person or entity seeking to have the work done. It may have the following entries:

- *Job or work order number* — Assigning a number to a permit request allows the work scheduler to keep track of who is working where.
- *Description of circuit/equipment/job location* — The requester must be specific and inclusive of all affected systems and their locations.
- *Description of the work to be done* — The requester must be specific as to what work is to be completed.
- *Justification of why the circuit or equipment cannot be de-energized or the work deferred until the next scheduled outage* — *NFPA 70E Section 110.2* discusses justifications for work involving electrical hazards. Thresholds for permitting exposure to electrical hazards can only be set by management. However, all workers have the right of refusal. If you are not capable, qualified, and comfortable with the work to be done, do not do it.

4.3.2 Completing Part II

Part II of the permit request identifies the procedures and safe work practices to be employed and must be completed by the electrically qualified people who will be doing the work. If there is a previously completed electrical risk assessment for the facility with both the shock risk assessment and arc flash risk assessment for the proposed work location, it can be used to complete the form. If current risk assessment data is not available or is suspect, the task must be delayed until the risk assessment is performed.

This part of the permit has numbered entries that must be checked off as they are completed. They are numbered sequentially to indicate that one action leads to the next, with the seventh item being signed off and dated by the electrically qualified persons. Attach extra pages as needed to fully describe the task(s) to be performed. The last entry of the part requires the actual qualified workers to sign off on the work.

4.3.3 Completing Part III

Part III of the example permit form is the authorization to perform work involving exposure to an electrical hazard. This section identifies everyone who may be responsible for or affected by the proposed work. Just as the qualified persons must believe the risk to be manageable from the perspective of injury, the appropriate level of management must also deem the risk to be acceptable to the employer. This involves considering what would happen in the event of an electrical incident where an appropriate risk assessment prevented employee injury but the equipment was damaged. An electrical incident will result in forced downtime or an outage. Even though the cost of repairing or replacing damaged equipment might be low, the possible price of schedule delays, data loss, process upset, and waste must be considered.

WARNING!

When properly selected, PPE reduces the risk of second-degree burns. It does not provide protection against physical trauma. See the Informational Note in *NFPA 70E Section 130.7(A)*.

Part III of the permit request requires the approval of the specified management personnel. All members of the team (including anyone engaged in perimeter security) must be a part of the task briefing and assist in completing the permit form.

Think About It

Permit Form

Why does the example permit found in *NFPA 70E Informative Annex J* have more lines for approvals than for qualified workers?

4.0.0 Section Review

1. The final step in a shock risk assessment is _____.
   a. identifying approach boundaries
   b. documenting the results
   c. determining any variables
   d. identifying normally energized components

2. The threshold incident energy level is the amount of energy required to produce _____.
   a. a second-degree burn
   b. holes or other openings
   c. an arc flash
   d. material breakdown

3. A situation that might require energized work is when a worker must perform operations on a _____.
   a. process control system
   b. grocery store refrigeration system
   c. life-support system
   d. street lighting system

5.0.0 Establishing Electrically Safe Work Conditions

Performance Tasks

There are no Performance Tasks in this section.

Objective

Explain how to establish electrically safe working conditions.
   a. Identify meters used to perform electrical testing.
   b. Explain lockout/tagout (LOTO) procedures.
   c. Describe emergency response procedures and personal safety requirements.

Specific procedures and test equipment are required to establish an ESWC. Note that these procedures may vary by location and jobsite. Always be aware of and follow the procedures in place at your jobsite.

5.1.0 Test Equipment

Know the Code

Test Instruments and Equipment
NFPA 70E Section 110.6

Know the Code

Testing
NFPA 70E Section 120.5(B)(6)

Various types of test equipment are used to establish an ESWC. Electrical test equipment is covered in *NFPA 70E Sections 110.6 and 120.5(B)(6)*. Test equipment falls into three general categories:

- *Daily use/tool pouch items* — This includes voltage sensors, voltage testers, continuity testers, multimeters, and clamp-on ammeters. Many meters combine several of these functions into one unit.

- *Specialty equipment* — This includes phase rotation testers, insulation resistance testers, portable power quality testers, and long-term metering equipment with associated test leads and components.

- *Safety equipment* — Safety equipment is designed to measure over a specific voltage range without requiring the use of a selector switch or input jacks.

All electrical test equipment must be inspected prior to use. Ensure the meter is properly rated for the circuit to be tested. Look for a cracked or oily case, broken input jacks, or any other deficiencies. Pay special attention to the test leads. Check that the test lead insulation is not cut, cracked, or melted, the tips are not loose, and the probes have finger guards.

Insulation testers or megohmmeters (*Figure 40*) must be inspected before each use to ensure they are in good working order. These meters are used only on de-energized circuits and equipment that have been verified to be in an ESWC. Hazardous energy levels are associated with these instruments, and PPE must be used per employer procedures.

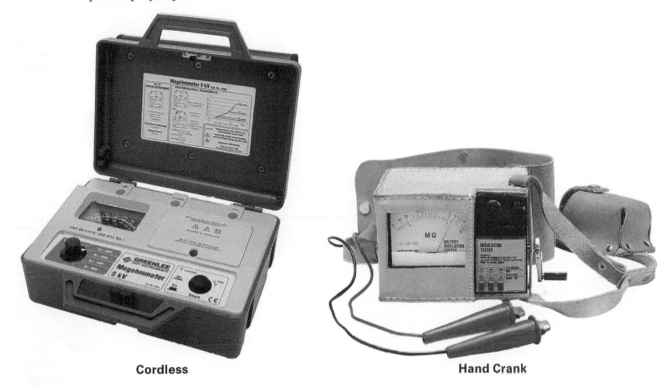

Cordless   **Hand Crank**

Figure 40 Megohmmeters.
Source: Ridge Tool Company

Proximity detectors such as voltage sensors are useful during the initial voltage survey of an area. The real value of voltage sensors is for the quick verification on known de-energized branch circuits before you touch a wire. However, they have the following limitations:

- Cannot discriminate voltage sources
- Cannot sense voltage through shielded cable
- May not detect voltage at some points on a twisted multiconductor cable
- May pick up too much background noise to be reliable
- Battery operated and will stop working without warning

Voltage detectors must be the contact type when verifying the absence of voltage in equipment such as switchgear and MCCs of the 1,000 V class. The detector selected should give a clear audible and visual indication of voltage presence. Most detectors of this type have a built-in self-test feature, eliminating the need for a high-voltage source. Noncontact voltage detectors are available in varying voltage ranges and, like any test equipment, cannot be relied upon if applied on systems outside of the listed range. They verify whether nominal phase voltage in the listed range is present.

WARNING!

Verifying the absence of voltage with a contact voltage detector may present the same shock and arc fault hazard potential as testing for voltage on an energized circuit. Before testing for the absence of voltage, always identify the electrical hazards present and select appropriate PPE for the anticipated level of both shock and arc hazards.

Medium- and high-voltage proximity detectors are suitable for most overhead applications as in substations or overhead lines. Remember, when verifying voltage in shielded cables, proximity detectors cannot be used. *Figure 41* shows a high-voltage proximity detector with an audio and visual voltage indicator.

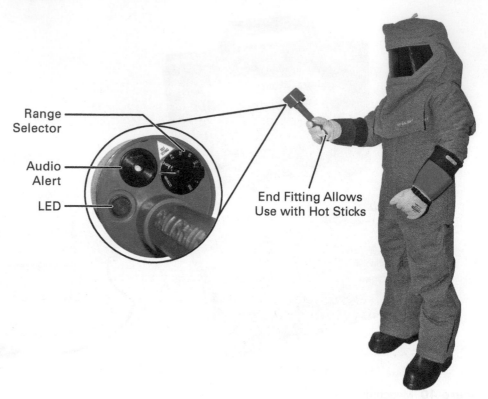

Range Selector

Audio Alert

LED

End Fitting Allows Use with Hot Sticks

Figure 41 Self-testing audio/visual voltage detector.
Source: Honeywell | Salisbury

Know the Code

Testing
NFPA 70E Section 110.6(A)

5.1.1 Training

NFPA 70E Section 110.6(A) requires that qualified persons be trained to select and use appropriate test instruments to verify the absence of voltage. This training must ensure the employee understands all limitations of each instrument that may be used. This means the employer is to identify voltage test instruments and the verification method to be used for each application that may be encountered.

5.1.2 Meter Use

Electronic multimeters are versatile tools used for diagnosis and troubleshooting. Many multimeters are auto-ranging and self-calibrating. They also have protective functions that prevent selector switch or test lead connection mistakes, such as performing a resistance test on an energized circuit.

Always measure voltage at the point of the lowest available energy. For example, if you are measuring voltage on a breaker panel, identify the lowest-rated breaker and make your measurement there. This way, you have more protection between yourself and the potential hazard. Take steps to obtain the best reading within the necessary envelope of safety. If conditions require that both of your hands remain free for a safe measurement, use an instrument stand or a magnetic hanger to hang the unit at eye level on the edge of the panel. Don't try to watch the meter while you move your test leads; always keep your eyes on the test probes.

WARNING!

All circuits are considered energized until proven otherwise.

When taking single-phase measurements, always connect the neutral lead first and the hot lead second. After taking your reading, disconnect the hot lead first and the grounded lead second. When testing for voltage, use the three-step test method:

Step 1 Test on a known energized circuit first.

Step 2 Test on the de-energized circuit.

Step 3 Retest on the first known energized circuit.

This process is critical to your personal safety and verifies that your test instrument is working properly.

There are many cases of electricians being injured or killed because they did not use the proper test equipment for the application. Test equipment and accessories are rated for both the voltage and current of the circuits and equipment on which they are to be used. Avoid the following common errors:

- Measuring voltage while test leads are in the current jacks
- Measuring AC voltage on the DC scale
- Contact with AC or DC circuits while in the resistance mode
- Prolonged contact with an energized circuit component
- Using a meter above its rated voltage
- Using a meter with damaged or improperly installed probes
- Not using a reliable ground, which will result in a voltage reading of 0 V phase-to-ground, even if voltage is present
- Using fuse end caps (ferrules) as test points, as some are insulated and will not show voltage

When working with energized circuits, use all appropriate PPE, including insulating gloves rated for the voltages present. The PPE worn should be based on the arc flash analysis and the incident energy available at the point of test.

Avoid holding a meter in your hands to minimize personal exposure to the effects of transients. Hang or rest the meter on a stand whenever possible. Many meters include magnetic attachments for hanging.

Some meters provide a wireless connection to smartphones (*Figure 42*). This can enhance safety as the electrician performing the test can focus on the placement of the probes while another worker records the readings. The use of smartphones also allows data logging and video conference calling to share live readings with a specialist at a remote location.

Make sure both the meter and leads have the correct category rating for each task, even if it means switching meters throughout the day. Test leads are an important component of digital meter safety. Look for test leads with double insulation, shrouded input connectors, finger guards (*Figure 43*), shielded tips, and a nonslip surface. This reduces the risk of shock and arc flash. In addition, single-function auto-detect safety voltmeters can be used to avoid using the wrong scale when taking measurements.

CAUTION

Do not change test settings between steps. Do not turn the instrument off and on between steps.

Safety Meters

Safety meters have touch-safe, nonremovable probes and leads, as well as shrouded tips. A safety meter provides an additional margin of safety for the qualified person.

Did You Know?

Solenoid-Type Voltage Testers

Solenoid-type voltage testers are increasingly prohibited in modern industrial plants and projects. Their simple coil construction has low impedance, which results in higher test currents. Higher currents make a significant arc during make-and-break contact. In addition, these testers may not indicate low voltages. Overheating of the coil can also be an issue, so the manufacturer's cooldown recommendations between voltage tests must be observed. Older solenoid testers are not marked with category ratings and therefore cannot be used in most of the situations encountered by an electrical worker.

Figure 42 Meter with wireless smartphone link.
Source: Tri-City Electrical Contractors Inc.

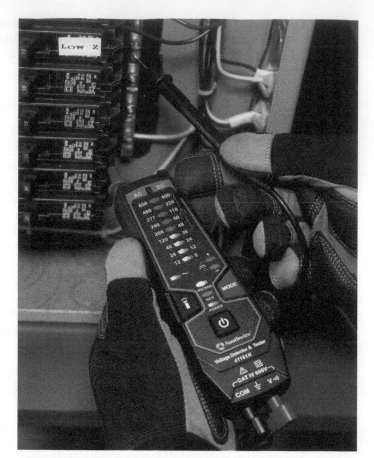

Figure 43 Basic voltage tester with finger guards.
Source: Southwire

Overvoltage installation category rating: The ability of an instrument to withstand voltage spikes up to a specified level. There are four category ratings (CAT I through CAT IV). In general, the higher the category rating, the greater the danger posed by voltage surges or transients.

UL/IEC Standard 61010-1, Safety Requirements for Electrical Equipment for Measurement, Control, and Laboratory Use - Part 1: General Requirements, addresses safety requirements for electrical equipment for measurement, control, and laboratory use. Electrical test equipment is marked with a specific **overvoltage installation category rating** (CAT I through CAT IV). In general, the higher the category number, the greater the danger posed by transients or voltage surges. These categories depend more on the fault current available at the point in the distribution system than the voltage level. Within each category, there are voltage ratings at 1,000 V, 600 V, and 300 V. *Table 3* shows overvoltage category ratings for various installations. These categories include the following:

- CAT I refers to protected electronic circuits.
- CAT II covers the receptacle circuit level and plug-in loads.
- CAT III covers distribution-level wiring. This includes 208 V, 480 V, and 600 V circuits, such as three-phase bus and feeder circuits, MCCs, load centers, and distribution panels. Permanently installed loads are also classified as CAT III. CAT III includes large loads that can generate their own transients. *Figure 44* shows a phase sequence tester with a CAT III rating.
- CAT IV is associated with the origin of installation. This refers to power lines at the utility connection, as well as the service entrance. It also includes outside overhead and underground cable runs since both may be affected by lightning.

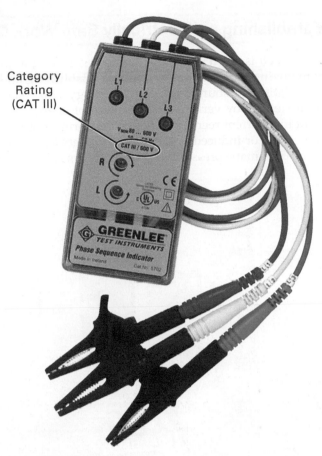

Figure 44 Meter category rating.
Source: Ridge Tool Company

TABLE 3 Overvoltage Installation Categories

Overvoltage Category	Installation Examples
CAT I	Electronic equipment and circuitry
CAT II	Single-phase loads such as small appliances and tools; outlets at more than 30' from a CAT III source or 60' from a CAT IV source
CAT III	Three-phase motors, single-phase commercial or industrial lighting, switchgear, and bus duct and feeders in industrial plants
CAT IV	Three-phase power at meter, service entrance, or utility connection; any outdoor conductors

All voltage testers should be considered safety equipment and included in the general category of PPE that receives special attention and care, including periodic calibration as required.

WARNING!

Any tool can be a hazard in the hands of an untrained operator. A complete understanding of the tool's characteristics, operation, and limitations is essential for personal safety. Be sure you are properly trained to use a specific meter before using it.

5.2.0 Establishing an Electrically Safe Work Condition

Lockout/tagout (LOTO) is the term often used to describe how to make equipment safe by removing the potential for hazard. Establishing an ESWC applies specifically to isolating the sources of electrical energy. Other sources of energy must also be made safe by verifying the absence of energy and taking measures to prevent inadvertent reenergization of the normal source. Other energy sources include static or induced voltages, gravity, tension, thermal, chemical, and hydraulic or pneumatic pressures. *Figure 45* shows LOTO devices.

(A) Electrical Lockout

(B) Pneumatic Lockout

Figure 45 LOTO devices.
Source: Panduit Corporation

Energized conductors and circuit parts an employee may be exposed to must be put into an ESWC before an employee works within the limited approach boundary, per *NFPA 70E Section 110.2*. *NFPA 70E Article 120* provides guidance on making a workspace safe from electrical hazards.

The steps in achieving an ESWC are identified in *NFPA 70E Section 120.5*. These are standard procedures for LOTO in nearly every safety policy. *NFPA 70E Section 120.5* is written specifically for an electrical lockout. The procedure is as follows:

Step 1 After donning the appropriate PPE, identify all sources of electrical supply. Check all applicable current drawings and tags.

Step 2 Shut down the equipment by normal means and open the disconnecting device(s).

Step 3 Visually verify the opening of switchblades where possible or full drawout of drawout-type circuit breakers (Figure 46).

WARNING!

Do not expose yourself to a potential shock or arc event without completing a risk assessment and energized work form.

Step 4 Release stored electrical energy.

Step 5 Release or block stored nonelectrical energy.

Step 6 Apply LOTO devices per documented employer policy.

Step 7 Verify the absence of voltage while wearing PPE that was selected based on the completed risk assessment. Test each phase conductor or component both phase-to-phase, phase-to-neutral (if available), and phase-to-ground. Before and after each test, check that the test instrument is operating properly by testing it on a known voltage source. When there is possibility of induced voltage or stored energy, such as capacitors, ground the conductors or circuit components before touch.

Step 8 When it can be anticipated that de-energized circuit parts could be reenergized, apply temporary grounds rated for the available fault current while wearing appropriate PPE.

Case History

Requesting an Outage

An electrical contractor requested an outage when asked to install two bolt-in, 240 V breakers in panels in a data processing room. It was denied due to the 24/7 worldwide information processing hosted by the facility. The contractor agreed to proceed only if the client signed a letter agreeing not to hold them responsible if an event occurred that damaged computers or resulted in the loss of data. No member of upper management would accept liability for this possibility, and the outage was scheduled.

The Bottom Line: If you can communicate the liability associated with an electrical event, you can influence management's decision to work on energized equipment.

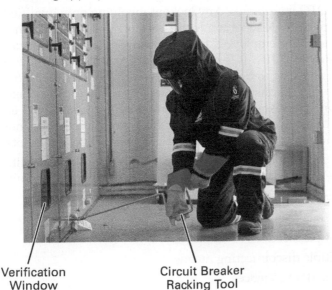

Verification   Circuit Breaker
Window   Racking Tool

Figure 46 Verification window.
Source: Oberon

Permanently Mounted Absence-of-Voltage Testers

Permanently mounted absence-of-voltage testers (AVTs) are now recognized by Exceptions No. 1 through 7 in *NFPA 70E Section 120.6(7)* as an alternative to the use of portable test equipment for verifying an ESWC. These panel-mounted pushbutton testers offer several safety advantages over the use of portable test equipment. For example, the use of a portable tester requires an open cabinet, worker exposure to hazards, and a multistep testing process (test the tester, check for voltage, then retest the tester). A permanently mounted AVT allows for testing in seconds with no exposure to potential electrical hazards within the cabinet.

Portable Tester
10 to 20 Minutes

Absence of Voltage Tester (AVT)
Less than 10 Seconds

Source: Panduit Corporation

5.2.1 Electrical LOTO

NFPA 70E Section 120.2(A) requires employers to identify, document, and implement LOTO procedures conforming to the requirements of *NFPA 70E Article 120*. *NFPA 70E Section 120.3* lists required principles for the execution of LOTO, including employee involvement, coordination, and control.

5.2.2 Hazardous Energy Control Procedures

There are two basic electrical energy control procedures allowed by *NFPA 70E Section 120.3(H)*:

- *Simple LOTO procedure* — Any lockout involving only qualified persons de-energizing a single set of conductors or electrical source for the sole purpose of safeguarding employees from exposure to electrical hazards is considered a simple LOTO. It is important to recognize that unqualified personnel may not work under this procedure. *OSHA Standard 29, Section 1910.147* identifies conditions for simple lockout procedures for nonelectrical work. Those procedures may be implemented by any authorized employee. The energy control procedures of your employer may or may not distinguish between simple lockout for electrical and simple lockout for other purposes.

- *Complex LOTO procedure* — When any of the following conditions are present, *NFPA 70E Section 120.5(A)(5)* requires a complex lockout procedure:
  - Multiple energy sources
  - Multiple crews or crafts
  - Multiple locations or employers
  - Multiple disconnecting means
  - Particular sequences
  - Multishift work

Know the Code

General
NFPA 70E Section 120.2(A)

Know the Code

Lockout/Tagout Principles
NFPA 70E Section 120.3

Know the Code

Forms of Control of Hazardous
Electrical Energy
NFPA 70E Section 120.3(H)

Know the Code

Complex Lockout/Tagout
NFPA 70E Section 120.5(A)(5)

Case History

Arc Flash Incident

An electrician was assigned to check out a processing line. The line had already been down for 15 minutes, and the production supervisor was anxious. After troubleshooting the system, the electrician determined that the motor starter to the main conveyor motor was not working. The fault protective devices on the starter would not reset. To save time, the electrician decided to replace the whole motor starter assembly (a common practice). By the time the electrician went to the shop, obtained a replacement assembly, and returned to the line, it had been down for 45 minutes. The production manager again asked when the line would be back up. The electrician opened the disconnect switch supplying electrical power to the starter, installed a lock, verified that the circuits were de-energized, replaced the motor starter assembly, and removed the lock on the disconnect switch. After quickly doing a visual inspection of all the wires and terminals, he closed the disconnect switch while the enclosure door was still open. Suddenly, there was a bright light, an intense flash of heat, and the electrician was thrown backward with severe burns and other injuries.

The Bottom Line: Never reenergize or reset equipment without first determining why the original component failed. Also, never let anyone rush you into losing sight of your own safety.

The intent of the detailed requirements of a complex lockout is to require a written execution plan and to assign specific responsibility and leadership. Logistics involving multiple sources to be locked out and/or multiple employees and crews who must place personal locks under this policy must be defined, and procedures for group lockout devices or use of lockboxes must be implemented.

Only with a written execution plan and written procedures can the complex LOTO procedure account for all persons who might be exposed per *NFPA 70E Section 120.5(A)(5)*. Each of these conditions is described in more detail in the following sections.

5.2.3 Multiple Energy Sources

Process lines often have machinery powered by electrical, hydraulic, and pneumatic systems. Many lines are designed so that the devices controlling various energy sources are all served by one master disconnect. Others require separate lockouts for each energy source.

5.2.4 Multiple Crews or Multiple Crafts

Complex lockouts ensure that all workers are protected equally. For example, it is common for a maintenance crew and a cleanup crew to be working on a line or machine at the same time. Anyone working in the area must participate in the LOTO.

5.2.5 Multiple Locations

Some systems have multiple locations where the electrical power to a machine can be disconnected. In most cases, the disconnect located closest to the machine or process is normally locked out, but if that point cannot be used, use the next upstream lockout point.

5.2.6 Multi-Shift Work

When a job or task continues for more than one work period or shift, the person in charge of a complex LOTO procedure is responsible for making sure that all personnel involved in completing the work are cleared at the end of the shift, and all locks are accounted for before turning the work over to the person in charge of the next shift.

Awareness Saves Lives

Increased awareness of workplace electrical hazards saves lives. Although fatal workplace electrical injuries has fallen in recent years, even one death is too many. Most electrical deaths are caused by contact with energized electrical equipment and wiring, which can be prevented by following appropriate LOTO procedures along with the other safe working requirements detailed in NFPA 70E®.

Source: *Electrical Safety Foundation International (ESFI)*

Know the Code

General Requirements for Electrical
Safety-Related Work Practices
NFPA 70E Article 110

Know the Code

Establishing an Electrically Safe
Work Condition
NFPA 70E Article 120

Know the Code

Elements of Work Permit
NFPA 70E Section 130.2(B)

Know the Code

Risk Assessment and Risk Control
NFPA 70E Informative Annex F

Know the Code

Job Briefing and Job Safety
Planning Checklist
NFPA 70E Informative Annex I

Know the Code

Other Precautions for Personnel
Activities
NFPA 70E Section 130.8

Know the Code

Changes in Scope
NFPA 70E Section 130.8(A)(3)

5.3.0 Personal Safety and Emergency Response

Following the requirements of *NFPA 70E Article 110* will help you recognize when an electrical hazard may be present. Making use of the principles, controls, and procedures of a comprehensive electrical safety program is the foundation of an effective safety culture within any organization. Combining the knowledge you have gained with a principled effort to be responsible for your personal safety and that of your coworkers will result in a safer work environment.

5.3.1 Personal Safety Toolbox

One of the most basic barriers to personal safety is poor communication. Relying on verbal communication can create unintended barriers. Communication must be clear and specific. When you receive instructions, verbally repeat them and ask for confirmation. If you are not sure of the intent of the instructions, ask. When giving directions, ask that they be repeated back to you, and if there are any questions, ask again until you are certain you are understood. People often agree on wording while having very different ideas of what those words mean.

Many opportunities for error can be eliminated by clearly identifying specific equipment by name in task instructions, paired with proper equipment labeling both on the operating face and at each removable cover on the back of the equipment. Marking look-alike equipment and equipment outside the work scope or boundaries also helps to eliminate error.

Performing lockout and creating an ESWC as described in *NFPA 70E Article 120*, together with a test-before-touch mentality, will remove most employees from exposure to electrical hazards.

When energized work must be performed, use of the Energized Electrical Work Permit process required by *NFPA 70E Section 130.2(B)*, the hazard/risk evaluation procedure of *NFPA 70E Informative Annex F*, and the job briefing and job safety planning checklist of *NFPA 70E Informative Annex I*, will result in effective planning, communication, and accountability for all parties.

NFPA 70E Section 130.8 identifies other precautions for personnel activities. These rules should be included in the employer's electrical safety program. All depend on the willingness of the employee to follow them, including recognizing when they might be tired, distracted, or otherwise impaired.

A worker's reaction to unexpected conditions or the need to perform additional work often means the difference between going home safely and causing an incident. When conditions are not per plan, you must back up and reassess the situation. See *NFPA 70E Section 130.8(A)(3)*. A large percentage of electrical injuries and fatalities involve changes in scope (scope creep), which may result in workers going outside of the boundaries of an established ESWC and bypassing a barrier or removing a cover to unknowingly access exposed energized parts.

5.3.2 Emergency Response Plan

Victims of an electrical event are injured by any combination of electrical shock, arc flash, and arc blast. At each instance when exposure to an electrical hazard is recognized, an emergency response to an incident must be planned. Many industrial facilities and large construction sites have an emergency response plan in place and practice it periodically. Smaller locations also may have emergency response plans in place. You must review these plans to supplement your response efforts. Often, emergency response professionals, such as fire, ambulance/EMTs, or air rescue, have a working relationship and plan of access for the facility. This saves valuable minutes during a rescue effort.

Communication is one of the most important parts of any emergency response plan. Take a moment at each jobsite and imagine how you would describe your location to a 911 operator in the event of an emergency. This is especially important for contractor personnel. Make sure someone is assigned to meet with emergency personnel in the event of an electrical incident. Emergency personnel need to be taken to the person who has been injured.

- *Shock* — When you get shocked, your survival depends entirely on the path the voltage takes through your body. A surface path may only produce burns at the entry and exit points, while a path across your heart might result in death. Victims of electrical shock may not be able to let go of the object(s) shocking them. Accordingly, *NFPA 70E Section 110.4(C)(1)* requires annual contact release training for employees responsible for safely releasing those who are unable to let go after making contact with energized conductors. Anyone trying to help them must remember not to touch them or the rescuer also becomes part of the circuit and a victim. If the source of the electricity cannot be turned off immediately, use an insulated rescue hook (*Figure 47*) to remove the person from the electrical source. Be ready to catch the victim once pulled off the source to prevent more injuries from occurring. After a victim has been removed, the area must be made safe for the rescuers so that first aid can begin. A victim of electrical shock may go into physical shock. Anyone trying to help the victim must understand how to treat shock victims. This is why *NFPA 70E Section 110.4(C)(2)* requires standby personnel to be trained in first aid, cardiopulmonary resuscitation (CPR), and the use of automated external defibrillators (AEDs). The heart of a shock victim may stop or lose its normal rhythm (known as *fibrillation*). If someone's heart has stopped, use CPR to maintain the victim's breathing and blood flow until help arrives. A heart in fibrillation can only be restored to normal rhythm using a defibrillator. AEDs are increasingly common in public areas, on airplanes, and in the workplace. An AED (*Figure 48*) can be used to check a victim's heart rhythm and indicate when or if a shock is required. These machines automatically sense, diagnose, and provide user guidance, requiring minimal training for effective use.

- *Arc flash* — The victim of an arc flash is often injured much more seriously than is first apparent. Douse the victim with water and remove them to a safe location. Be aware that burns to the neck may interfere with the victim's ability to breathe. Most arc flash victims incur injuries so severe that they require treatment at a special burn center.

While most companies now have emergency response teams to react to any safety emergency, a safety team that is not present cannot react quickly enough to save someone involved in an electrical event. When performing work on energized equipment, always have a trained person on standby to help if needed. *OSHA Standard 29, Section 1910.269* requires that CPR be available within four minutes where employees may be exposed to a shock hazard. *NFPA 70E Section 110.4(C)* identifies training requirements for employees who may be exposed to shock hazards. Training includes first aid and emergency response, as well as methods of releasing shock victims from electrical contact.

Electricians may have to work in remote areas that are off-limits to other employees. These may include switchgear rooms and areas housing MCCs. Anyone working in these areas must ensure supervisors, managers, and coworkers know where they are and what they are doing. This is a good example of why Energized Electrical Work Permits are required.

Think About It

Eliminating Electrical Incidents

How many days have you worked without injury? What about your crew or your jobsite? Was the last injury preventable? What can you do to prevent a recurrence of that event?

Know the Code

Emergency Response Training
NFPA 70E Section 110.4(C)

Think About It

Calling for Help

Does your hazard assessment include the name and address of the facility you are working at for anyone who may need to call for first responders?

Figure 47 Insulated rescue hook.
Source: Honeywell | Salisbury

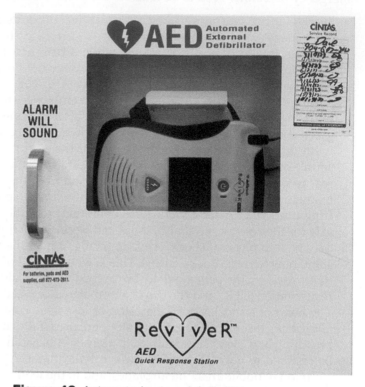

Figure 48 Automated external defibrillator.

5.0.0 Section Review

1. The ability of an instrument to withstand voltage spikes up to a specified level is known as the _____.
   a. arc flash rating
   b. voltage rating
   c. short circuit withstand rating
   d. overvoltage installation category rating

2. Where in NFPA 70E® would you find an electrical lockout procedure?
   a. *NFPA 70E Section 110.3*
   b. *NFPA 70E Section 120.5*
   c. *NFPA 70E Section 130.1*
   d. *NFPA 70E Section 240.2*

3. In all locations where employees may be exposed to a shock hazard, *OSHA Standard 29, Section 1910.269* requires that CPR be available within _____.
   a. two minutes
   b. four minutes
   c. six minutes
   d. eight minutes

1. A shock of 5 mA can cause injury through _____.
   a. respiratory arrest
   b. suffocation
   c. involuntary movement
   d. electrical burns

2. The voltage between the feet of a person standing near an energized, grounded object is known as _____.
   a. touch potential
   b. electromotive potential
   c. shock potential
   d. step potential

3. Electrical equipment having a dead front is _____.
   a. unable to sustain an electrical arc
   b. free from shock hazard as long as doors and covers are in place
   c. classified as guarded equipment
   d. required to be tested annually for insulating capability

4. The *least* effective method of preventing risk is through the use of _____.
   a. elimination
   b. substitution
   c. engineering controls
   d. PPE

5. General requirements for safety-related work practices are found in _____.
   a. *NFPA 70E Article 90*
   b. *NFPA 70E Article 110*
   c. *NFPA 70E Article 115*
   d. *NFPA 70E Informative Annex C*

6. The approach limit within which a person could receive a second-degree burn if an electrical arc flash were to occur is known as the _____.
   a. limited approach distance
   b. shock protection boundary
   c. AFB
   d. restricted approach boundary

7. In order to cross a limited approach boundary, an unqualified person must be escorted by a(n) _____.
   a. manager
   b. OSHA compliance safety and health officer
   c. accountability partner
   d. qualified person

8. Job safety planning and a job briefing are required before starting any work involving _____.
   a. an ESWC
   b. exposure to electrical hazards
   c. unqualified persons
   d. LOTO procedures

9. Common errors that can occur in skill-based mode are due to _____.
   a. a perceived reduction in risk
   b. an inaccurate assessment of the situation
   c. deviating from an approved procedure
   d. applying the correct procedure to the wrong situation

10. Situations that put a worker at risk due to the demands of the task, conditions, worker attitude, and/or environment are known as _____.
   a. human performance modes
   b. error precursors
   c. potential risk points
   d. error flags

11. A calorie is equal to about _____.
   a. 1 joule
   b. 4 joules
   c. 15 joules
   d. 40 joules

12. Incident energy is measured in _____.
   a. volt/cm^2
   b. cal/cm^2
   c. amp-hours
   d. kWh

13. Requirements for inspection, care, and maintenance of arc-rated clothing and arc flash suits are found in _____.
   a. *NFPA 70E Section 110.5(B)(1)*
   b. *NFPA 70E Section 130.7(C)(13)*
   c. *NFPA 70E Section 250.2(B)(4)*
   d. *NFPA 70E Informative Annex F*

14. The color code for a Class 0 glove is _____.
   a. beige
   b. white
   c. red
   d. green

15. One attribute of a qualified person is that they _____.
   a. are knowledgeable of the construction and operation of the equipment involved
   b. carry a valid electrical license
   c. own required PPE
   d. have completed an electrical apprenticeship

16. Working distances associated with low-voltage and medium-voltage equipment can be calculated using the equations found in _____.
   a. *NFPA 70E Table 130.7(C)(14)*
   b. *NFPA 70E Tables 130.4(D)(a) and (b)*
   c. *NFPA 70E Informative Annex D*
   d. *NFPA 70E Informative Annex F*

17. Which of the following is *not* determined in an arc flash risk assessment?
   a. The AFB
   b. Appropriate PPE
   c. When an arc flash hazard is expected to exist
   d. The limited approach boundary

18. *NFPA 70E Table 130.7(C)(15)(a)* may be used to identify the PPE category for a task when _____.
   a. the available short circuit current is below given ranges
   b. both the available fault current and clearing time are below given values
   c. either the available fault current or the clearing time are below given values
   d. no information about the electrical system is available

19. Thresholds for permitting exposure to electrical hazards can only be set by _____.
    a. OSHA
    b. NFPA
    c. workers
    d. management

20. One advantage of a proximity detector is _____.
    a. quick noncontact verification of de-energized branch circuits
    b. background noise does not interfere with readings
    c. the ability to discriminate between voltage sources
    d. it can verify voltage along the full length of twisted multiconductor cable

21. The overvoltage installation category that covers electronic equipment and circuitry is _____.
    a. CAT I
    b. CAT II
    c. CAT III
    d. CAT IV

22. A simple LOTO procedure is *most* likely to be used with _____.
    a. work that involves more than one energy source
    b. multiple locations
    c. work that is separated into shifts
    d. a single set of conductors

23. A job briefing, and job safety planning checklist can be found in _____.
    a. *NFPA 70E Informative Annex A*
    b. *NFPA 70E Informative Annex C*
    c. *NFPA 70E Informative Annex F*
    d. *NFPA 70E Informative Annex I*

24. A large percentage of electrical incidents are caused by changes in _____.
    a. shifts
    b. temperature
    c. management
    d. scope

25. Which of the following is *true* regarding emergency response to an electrical shock?
    a. An AED may only be used by specially trained emergency response personnel.
    b. A heart in fibrillation can be restored to normal rhythm using CPR.
    c. Electrical shock always stops the heart.
    d. A heart in fibrillation can only be restored to normal rhythm using a defibrillator.

Answers to odd-numbered questions are found in the Review Question Answer Keys at the back of this book.

Answers to Section Review Questions

Answer	Section Reference	Objective
Section 1.0.0		
1. d	1.1.1; *Table 1*	1a
2. b	1.2.1	1b
3. b	1.3.3	1c
Section 2.0.0		
1. b	2.1.0	2a
2. a	2.2.1	2b
3. d	2.3.0	2c
Section 3.0.0		
1. d	3.1.0	3a
2. b	3.2.0	3b
3. a	3.3.0	3c
4. d	3.4.1	3d
Section 4.0.0		
1. b	4.1.3	4a
2. a	4.2.1	4b
3. c	4.3.0	4c
Section 5.0.0		
1. d	5.1.2	5a
2. b	5.2.0	5b
3. b	5.3.2	5c

User Update

Did you find an error? Submit a correction by visiting **https://www.nccer.org/olf** or by scanning the QR code using your mobile device.

SCAN ME

REVIEW QUESTION ANSWER KEYS

MODULE 26501-24

Answer	Section Reference
1. c	1.1.1; *Table 1*
3. b	1.2.4
5. b	2.0.0
7. d	2.1.2
9. a	2.3.1
11. b	3.0.0
13. b	3.3.1
15. a	4.1.1
17. d	4.2.1
19. d	4.3.1
21. a	5.1.2; *Table 3*
23. d	5.3.1
25. d	5.3.2

ADDITIONAL RESOURCES

26501-24 Managing Electrical Hazards

IEEE 1584-2018, *IEEE Guide for Performing Arc Flash Hazard Calculations*. 2018. New York, NY: Institute of Electrical and Electronics Engineers.

NFPA 70®-2023, *National Electrical Code®*. 2023. Quincy, MA: National Fire Protection Association.

NFPA 70B-2023, *Standard for Electrical Equipment Maintenance*. 2023. Quincy, MA: National Fire Protection Association.

NFPA 70E®-2024, *Standard for Electrical Safety in the Workplace®*. 2024. Quincy, MA: National Fire Protection Association.

GLOSSARY

Arc blast: The explosive expansion of air and metal in an arc path. Arc blasts are characterized by the release of a high-pressure wave accompanied by shrapnel, molten metal, and deafening sound levels.

Arc fault: A high-energy discharge between two or more energized conductors or an energized conductor and ground.

Arc flash: A dangerous condition caused by the enormous release of thermal energy in an electric arc, usually associated with electrical distribution equipment.

Arc flash boundary (AFB): An approach limit at a distance from exposed energized electrical conductors or circuit parts within which a person could receive a second-degree burn if an electrical arc flash were to occur.

Arc flash hazard: A dangerous condition associated with the release of energy caused by an electric arc.

Arc flash risk assessment: A study investigating a worker's potential exposure to arc flash energy, conducted for the purpose of injury prevention, determination of safe work practices, and appropriate levels of PPE.

Arc rating: The maximum incident energy resistance demonstrated by a material (or a layered system of materials) prior to material breakdown or at the onset of a second-degree skin burn. Expressed in J/cm^2 or cal/cm^2.

Arc thermal performance value (ATPV): The incident energy limit that a flame-resistant material can withstand before it breaks down and loses its ability to protect the wearer. Expressed in J/cm^2 or cal/cm^2.

Bolted fault: A short-circuit or electrical contact between two conductors at different potentials in which the impedance or resistance between the conductors is essentially zero.

Calorie: The amount of heat energy required to raise the temperature of 1 gram of water by 1°C.

Dead front: Equipment that has no exposed energized electrical conductors or circuit parts on the operating side.

Electrically safe work condition (ESWC): A state in which the conductor or circuit part to be worked on or near has been disconnected from energized parts, locked/tagged in accordance with established standards, tested to ensure the absence of voltage, and grounded if necessary.

Electrical shock: Occurs when a person or object is grounded and contacts another energized object. The sensation of being shocked occurs when current flows through tissues in the body.

Energy breakopen threshold (EBT): The incident energy limit that a flame-resistant material can withstand before the formation of one or more holes that would allow flames to penetrate the material. Expressed in J/cm^2 or cal/cm^2.

Error precursors: Situations that increase risk to a worker due to demands of the task, environmental conditions, and/or human error.

Incident energy: The amount of thermal energy impressed on a surface at a certain distance from the source of an electrical arc. Incident energy is typically expressed in cal/cm^2.

Limited approach boundary: An approach limit at a distance from an exposed energized electrical conductor or circuit part within which an electrical shock hazard exists.

Overvoltage installation category rating: The ability of an instrument to withstand voltage spikes up to a specified level. There are four category ratings (CAT I through CAT IV). In general, the higher the category rating, the greater the danger posed by voltage surges or transients.

Qualified person: A person who has the necessary training or certifications and has demonstrated the knowledge and ability to safely install and operate electrical equipment.

Restricted approach boundary: An approach limit at a distance from an exposed energized electrical conductor or circuit part within which there is an increased likelihood of electric shock.

Shock hazard: A dangerous condition associated with the possible release of energy caused by contact or approach to energized electrical conductors or circuit parts.

Step potential: The voltage between the feet of a person standing near an energized, grounded object, equal to the difference in voltages between each foot and the electrode.

Threshold incident energy level: The amount of energy required to produce a second-degree burn (1.2 cal/cm^2 or 5 J/cm^2).

Touch potential: The voltage between the energized object being touched and the feet of the person in contact with it, equal to the difference in voltage between the object and the grounding point.

Unqualified person: A person who does not possess the skills, knowledge, or training necessary to be a qualified person.